GIGABIT NETWORK

Advanced Information Technology

Series Editor
Tadao Saito

Volume 1

ISSN 1348-513X

Gigabit Network

Edited by

Tadao Saito
Chuo University, Japan

and

Hiroshi Esaki
The University of Tokyo, Japan

Tokyo • Amsterdam • Berlin • Oxford • Washington, DC

Gigabit Network

ISBN 4-274-90596-9 C3055 (Ohmsha)
ISBN 1-58603-351-4 (IOS Press)
Library of Congress Control Number: 2003106825

Publisher
Ohmsha, Ltd.
3-1 Kanda Nishiki-cho
Chiyoda-ku, Tokyo 101-8460
Japan

Distributor

USA and Canada
IOS Press, Inc.
5795-G Burke Centre Parkway
Burke, VA 22015
USA
fax: +1 703 323 3668
e-mail: iosbooks@iospress.com

UK and Ireland
IOS Press/Lavis Marketing
73 Lime Walk
Headington, Oxford OX3 7AD
England
fax: +44 1865 75 0079

Germany, Austria and Switzerland
IOS Press/LSL.de
Gerichtsweg 28
D-04103 Leipzig
Germany
fax: +49 341 995 4255

Japan
Ohmsha, Ltd.
3-1 Kanda Nishiki-cho
Chiyoda-ku, Tokyo 101-8460
Japan
fax: +81 3 3293 6224
e-mail: hanbaibu@ohmsha.co.jp
url: http://www.ohmsha.co.jp/index_e.htm

Netherlands and the rest of the World
IOS Press
Nieuwe Hemweg 6B
1013 BG Amsterdam
The Netherlands
fax: +31 20 620 3419
e-mail: order@iospress.nl
url: http://www.iospress.nl

Far East jointly by
Ohmsha, Ltd. and
IOS Press

v

Series Editor's Foreword

Information Processing Society of Japan (IPSJ) is the top academic institution in the information technology field in Japan. It has about thirty thousand members and promotes a variety of research and development activities covering all aspects of information technology.

One of the major activities of the society is publication of its transactions containing papers covering all the fields of information technology, including fundamentals, software, hardware, and applications. Some of the papers are published in English, but because the majority of the papers are in Japanese, the transactions are not suitable for non-Japanese readers wishing to access information technology in Japan. IPSJ therefore decided to publish a book series titled "Advanced Information Processing Technology" to facilitate access to Japanese information in the field of information technology. The first book in the series was published by Gordon and Breach, Science Publishers Inc. in 1998, and six books were published by 2002. The publisher was changed to Taylor & Francis in 2002. In 2003, the publisher was again changed to Ohmsha and IOS Press and the series title was changed to "Advanced Information Technology".

The series consists of independent books, each including top quality papers, from mainly Japanese sources in a selected area of information technology. The titles of books are selected by the International Publication Committee of IPSJ, to enable easy access to information for international readers. Each volume contains original papers and/or papers updated from original papers appearing in the IPSJ transactions or internationally qualified meetings. Survey papers to help understanding the overall situation in a particular area are also included.

As the chairman of the International Publication Committee of IPSJ, I sincerely hope that the books in the series will improve communication between Japanese and non-Japanese specialists for their mutual benefit.

Tadao Saito
Series Editor
Chairman
International Publication Committee
Information Processing Society of Japan

Preface

The digital information and communication infrastructure as empowered by Internet technology has already become an indispensable platform for people's daily lives, activities and working environment. The information and communication networks that had been based on voice communication infrastructure until the late 1990's could not provide a broadband, always-on communication environment to end users, and the BISDN system using ATM technology which had been developed in the second half of the 1980's could not serve as a widely-used digital information infrastructure. Internet technology, on the other hand, which utilized existing telephone infrastructure, has been developed with a significant growth rate in the 1990's, and recently a new type of information infrastructure using the native Internet technology such as ADSL and wireless LAN has begun to be deployed rapidly. As a result, it can now be said that Internet technology has secured its role as the key technology behind digital information systems in the 21st century. Internet technology has grown from a network solely for scientists which had been built on the existing telephone infrastructure into a necessary and critical part of the infrastructure supporting ordinary people's lives and activities, including many people who may know nothing about computers. The technology is now mutating toward a new stage, which is a broadband, native Internet environment.

When the Japan Gigabit Network (JGN) began operation in April 1999, most Internet connections used over 64Kbps class narrowband dial-up connections. Thus, the information and communication infrastructure was primarily a narrowband system and today's broadband environment had not yet been explored. It was under these circumstances that the JGN was initiated to create an environment enabling research and development of the infrastructure, middleware, and application technologies that would become necessary when a gigabit class network environment was introduced, and to accelerate this research and development. The typical pattern of evolution in Internet technology is that elemental technologies become integrated into an interactive Internet architecture. This environment then provides a real-world testing and evaluation by the general public users as well as by the architects of the Internet. This interaction ultimately provides a feedback to the R&D community that cycle back into the system itself. This particular feedback loop derives a creative and practical R&D effort on the Internet technology. In this sense, the R&D methodologies lying at the heart of the

JGN are very characteristic of the Internet. The same thing could be said of the focus of the R&D effort itself, which in fiscal year 2001 began work on IPv6 and which responds flexibly to internal and external technological trends while valuing feedback from society and industry for the determination of the project's general direction.

This book gives an overview of the R&D activities of the 7 projects utilizing the JGN, which accommodate the greatest number of participating organizations, and an overview of R&D testbeds in Europe, the United States and in Asia. Using these overviews, this book will show that the JGN project has borne many fruits, both domestic and international, academic and industrial. The source of these successes lies in the fact that the JGN provided an environment where the results of researchers' efforts are tested and evaluated in actual and practical use, and that the architecture and operation of this environment was in turn performed directly by the researchers themselves. Thus, the continuous operation of a new IT infrastructure on a trial basis has proved an exceptionally valuable policy for the promotion of Japan's information and communication technology research.

Finally, we would like to take this opportunity to express our gratitude to the many people who aided us in the editing of this book. In particular, we would like to thank Editor-in-Chief Dr. Eichi Wada and Ms. Wataya of the editing bureau for their astute comments and revisions of each of these articles. Without their gracious support and contributions, this publication would not have been accomplished.

April 3, 2003

Tadao Saito
Hiroshi Esaki

Contents

Gigabit Network
T. Saito and H. Esaki (Eds.)
Ohmsha/IOS Press, 2003

Chapter 1

A Summary of Japan Gigabit Network

Tadao Saito

Chuo University

Abstract. The JGN (Japan Gigabit Network) is an experimental network infrastructure for research and development toward Japan's first true next generation Internet. It has garnered worldwide attention in the more than 3 years since it began operation in April, 1999. At present, a variety of research activities are carried out on the JGN which have produced significant research results. In this article, I describe the purpose and background behind the JGN's system design, construction, network architecture, and R&D methodologies surrounding it. Finally, I give a brief overview of current research.

1. Background and Overview

At the end of the 1990's, R&D for very high-speed networks and next generation Internet technology was initiated via operation of vBNS and Abilene in the USA. Both of these testbeds are experimental infrastructures to advance R&D on high-speed networking technologies. Through these project activities, we realized that both the legacy communication infrastructure and network operations were technically incapable of responding to the needs of an emerging Internet which has been continuously growing in terms of both speed and size. Thus, we realized the urgent need for research to be carried out on next generation network technology utilizing an experimental gigabit class network. In 1998, the Telecommunications Advancement Organization (TAO) used supplementary budget appropriations to construct the JGN, which started its operation in fiscal year 1999. Since that time, JGN has worked to improve information and telecommunications technologies such as high-speed networking and high performance applications.

The JGN consists of ten ATM switches and fifty six nodes throughout Japan connected by an fiber-optic network (referred to as "gigabit network" below), five Shared Use Research Facilities, and Research Centers. The gigabit network and Shared Use Research Facilities are widely available for use in various R&D activities, including high-speed networking and high performance application technologies. The Research Centers are for gigabit network R&D activities directly under control of the TAO. The structure of the JGN is R&D centered on the gigabit network itself and its attendant facilities, with peripheral research on high-speed networking and high performance application technologies also proceeding apace.

The JGN is available for research use to universities, research organizations, governmental organizations, local government, and private companies. It is an open network utilized by a wide range of users. In order to support the establishment and deployment of the next generation Internet Protocol (IPv6), the JGN was upgraded so as to be IPv6 compliant in October of 2001, making it the largest native IPv6 network in the world. Through constant and rapid response to changes in information and communication technology, the JGN aims to be the world-wide center for R&D of very high-speed network technology.

2. JGN Network Structure

The JGN consists of three layers: Layer 1 (physical layer) with fiber-optics, (2) Layer 2 (data link layer) with ATM (Asynchronous Transfer Mode) technology, and Layer 3 (network layer) with IP (Internet Protocol). Layer 3 IP services includes both the widely used IPv4 and IPv6 (Internet Protocol Version 6), which is the next generation Internet protocol.

2.1 Physical Structure of the Gigabit Network (Layer 1 topology)

The physical structure of the JGN is shown in Figure 1. As shown in the map, the footprint of the JGN covers the whole of Japan; from Hokkaido in the north to Kyushu and Okinawa in the south.

Characteristics of transmission delays on the JGN is publicly available through the following URL: http://www.jgn.tao.go.jp/org_tec/chienzu.html

2.2 Structure of the ATM Network (Layer 2 topology)

The structure of the JGN's layer 2 ATM network is shown in Figure 2. User organizations can connect to JGN through ATM switches at sixty four locations. These access points cover all prefectures. As with the Layer 1 topology, the backbone covers the area from Sapporo in Hokkaido to Hakata in Kyushu with 2.4Gbps links, while other sites are connected with 50–600Mbps ATM links, most of which are connected with 150Mbps ATM

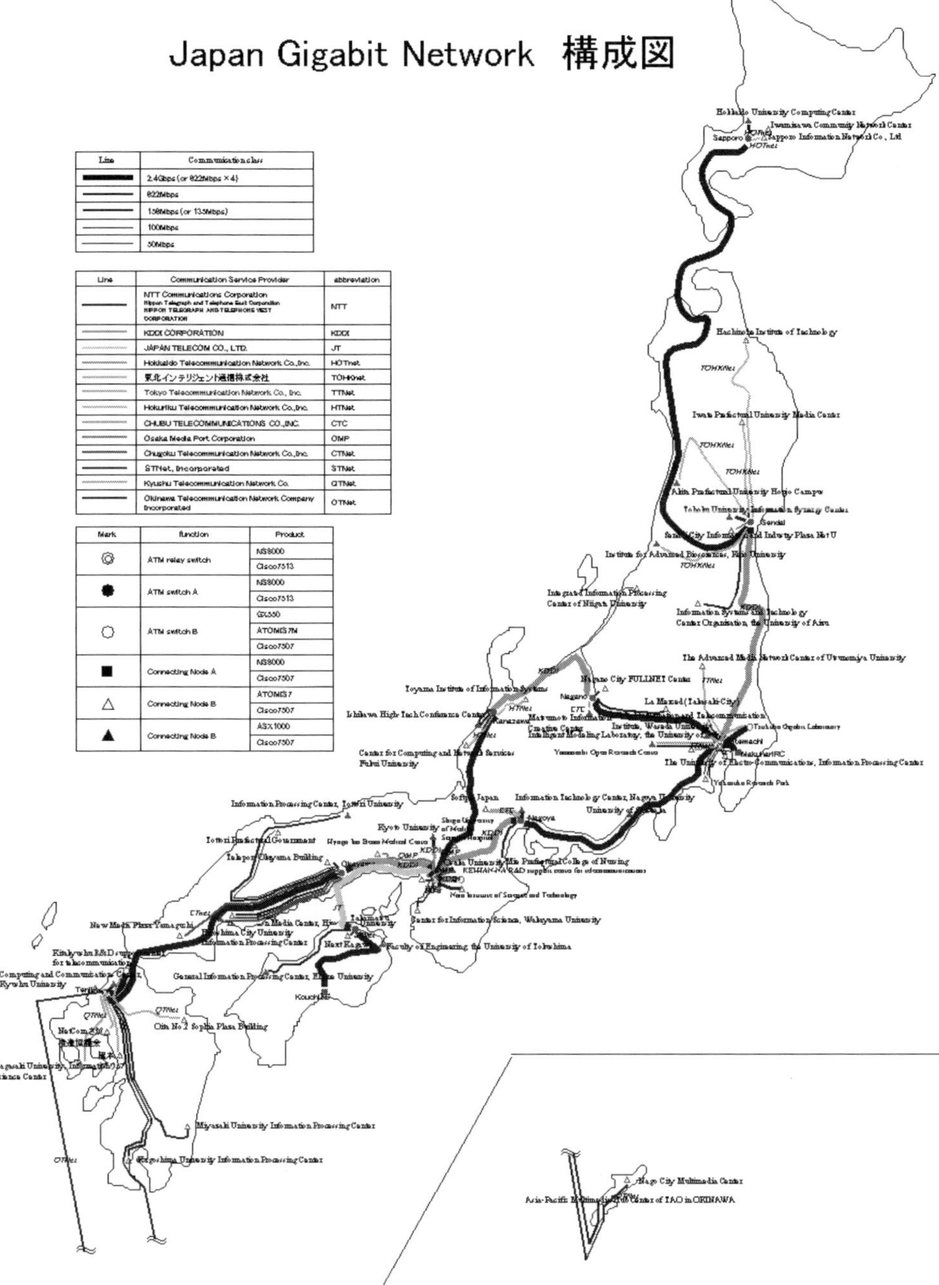

Fig. 1. Physical topology of Japan Gigabit Network.

links (OC-3). Users establish their own connections to the nearest JGN access point, most of which use OC-3 ATM SM (single mode) or MM (multimode) connections.

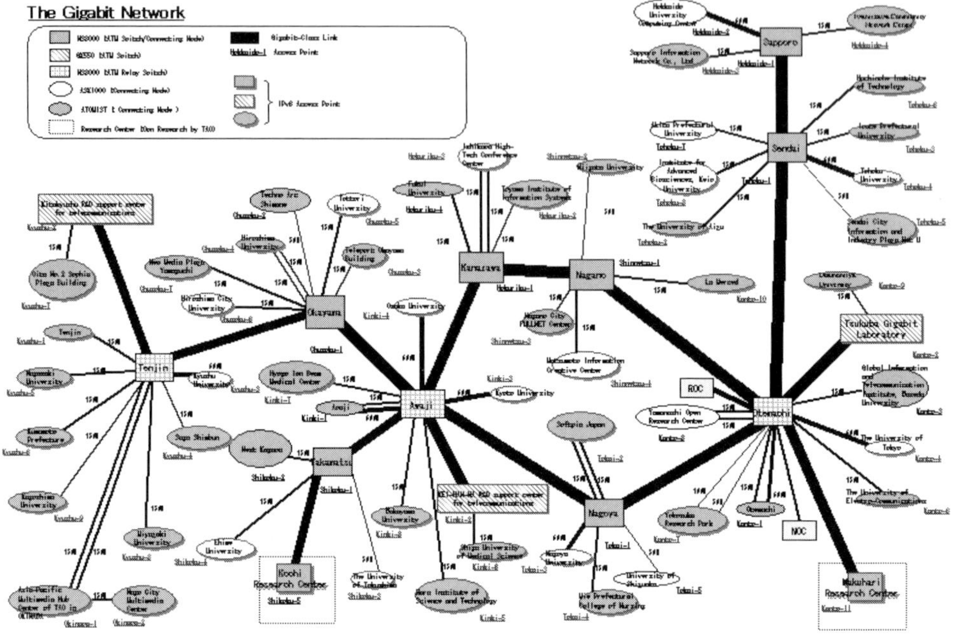

Fig. 2. Structure of JGN layer 2 ATM Network.

2.3 Structure of the IPv6 Network (Layer 3)

The structure of the JGN's IPv6 network is shown in Figure 3. JGN's IPv6 network has four core sites: the Okayama Interoperability and Evaluation Lab, the Computing and Communications Center of Kyushu University, and NTT Dojima in Osaka. Connectivity to user organizations is offered at other sites through routers and bridges. Finally, forty nine access points throughout the country are IPv6 compliant.

3. Promoting Research and Development

The JGN consists of two internal structures: one devoted to performing R&D and the other to promoting usage of the JGN (See Figures 4 and 5) [1].

3.1 Performing R&D

The internal structure within the JGN devoted to performing R&D is shown in Figure 4. This consists of 4 main groups: Research Centers performing research directly under control of the TAO, Gigabit Laboratories, publicly commissioned research, and general use research. R&D undertaken

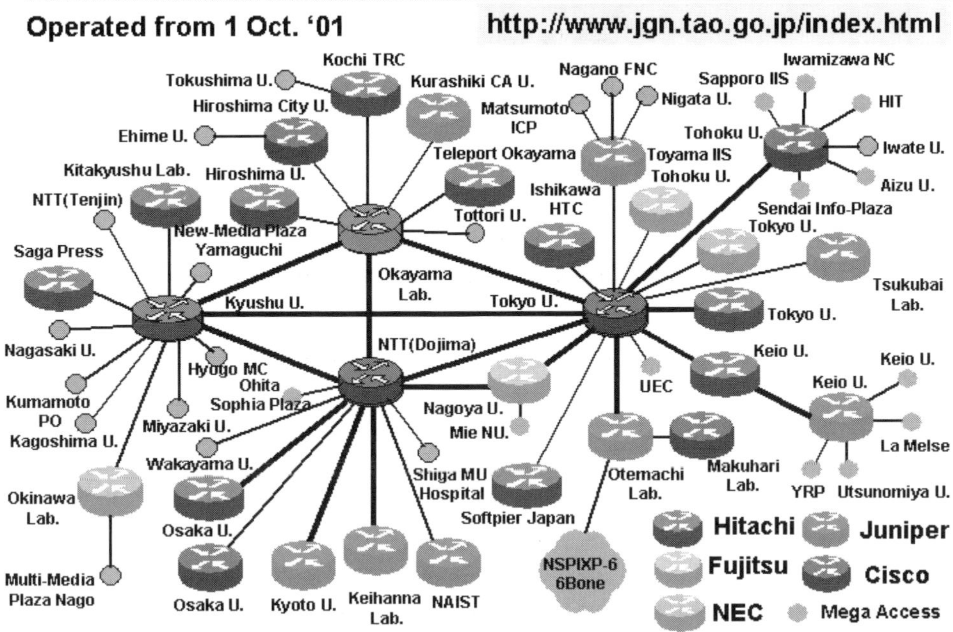

Fig. 3. Structure of JGN IPv6 Network.

Gigabit Network R&D Structure

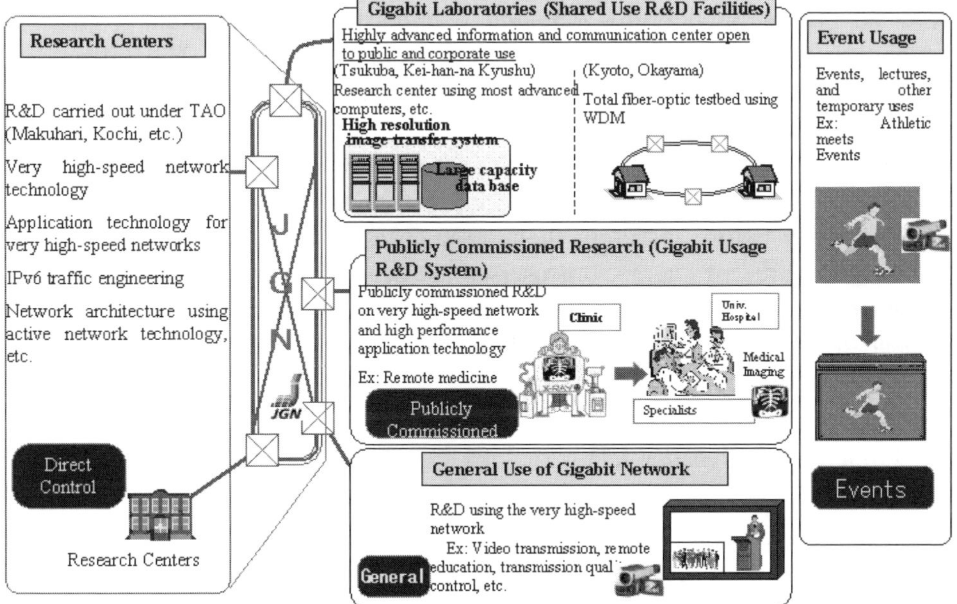

Fig. 4. Japan Gigabit Network's R&D system.

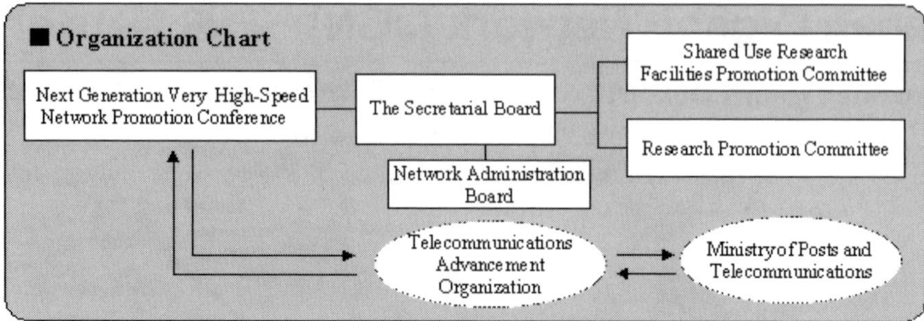

Fig. 5. Next generation very high-speed network promotion.

at research centers focuses on communication and transmission mechanisms, which is called "directly controlled research." Direct control Research Centers perform R&D on high-speed networking and high performance application technologies, IPv6 traffic engineering, and network architecture using technologies such as active networks.

Gigabit Laboratories are Shared Use Research Facilities in 5 locations throughout the country: Tsukuba, Keihanna, Kyoto, Okayama, and Kita-Kyushu. Each of these facilities contains ATM switches, IPv6 routers, visual equipment, and other state-of-the art resources, all of which are used freely by researchers.

Publicly commissioned research is open to public and private universities, national research organizations, national research institutions, private corporations, third sector institutions, and public-service corporations. The TAO advertises publicly for research commissions for research projects to which any of the above types of organizations may apply. The budget for these projects is limited to about 20 million yen per year, with a maximum project length of 3 years.

General use research is the most widely conducted form of research on the JGN. Those who wish to utilize the network sign a shared research contract with the TAO before initiating their research activity using JGN. Use of the JGN itself is free of charge, but users must bear all expenses for establishing a connection to the nearest access point. Finally, all users must agree to and respect the AUP (Acceptable User Policy), which is described later.

3.2 Promoting Use of the JGN

The Next Generation Very High-Speed Network Promotion Conference has been established in order to tackle issues such as promoting use of the JGN and solving various operation-related problems (Figure 5). The purpose of this organization is promoting research on the JGN and ensuring its smooth and efficient operation. The Conference meets once per year, and its fourth meeting was March 6, 2002. Further, the Secretariat Board, the Research Promotion Committee, the Shared Use Research Facilities Promotion Committee, and the Network Administration Board work under the auspices of the Next Generation Very High-Speed Network Promotion Conference. Their functions are described below.

The Secretariat Board is a conference devoted to ensuring the smooth operation of the Next Generation Very High-Speed Network Promotion Conference as a whole. The Research Promotion Committee shares beneficial information about the JGN and promotes its use and research on it. This committee is composed of users of the JGN. The Shared Use Research Facilities Promotion Committee works toward the effective use of the Shared Use Research Facilities. Finally, the Network Administration Board studies network operation policy from a technical viewpoint and makes suggestions about communications and transmission mechanisms in order to achieve smooth and efficient operation of the network.

Out of all these organizations, the Network Administration Board meets the most frequently, generally on a bimonthly schedule, where there are active technical discussions about JGN operation policy.

4. State of Research and Development

Below we will describe the state of R&D on the JGN using public data from the end of July, 2002.

4.1 Directly Controlled Research

The TAO utilizes the JGN to perform directly controlled research, formally called "Research and Development On Next Generation Very High-Speed Large Scale Networks". This embodies four areas of research, which are described briefly below. Directly controlled research is also described in more detail in the "Directly Controlled TAO Research Using the JGN (TAO Research Centers at JGN's Birth)" and "The JGN IPv6 Network" in this volume.

(1) Research on operation technologies and other aspects of very high-speed large scale networks

(2) High performance applications and research on virtual reality communication formats

These two areas of research are centered mainly at the Makuhari Gigabit Research Center, the University of Tokyo, the Kochi Communication

Traffic Research Center, and Tohoku University. The Makuhari Gigabit Research Center and the University of Tokyo are working on topics such as architecture environments for various applications, researching effective ways to use networks for applications with different characteristics, high precision network performance measuring methods, QoS (Quality of Service) implementation methods, and the creation of new applications suited to the gigabit network. Meanwhile, the Kochi Communication Traffic Research Center is conducting research on improving QoS, systems for high-speed and efficient parallel decoding of video sent from multiple locations, virtual reality communications systems enabling conversations and 3D images of participants, virtual architecture and communications formats which will spur demands for gigabit class networks, television-on-demand formats maintaining simultaneous streams, and other topics.

Tohoku University is investigating methods to collect and analyze network information, flexible QoS which is centered on the NIWH (Network Information Warehouse) that will present this data, "JaNI" (Japan Gigabit Network Information System), which will display traffic on the JGN in real time in a graphical format, storage systems for high-speed image systems, distributed network management systems, and other topics.

(3) R&D of IPv6 Traffic Engineering Technologies

Research in this area is undertaken by the Otemachi IPv6 System Operation and Technical Center, the Okayama Interoperability and Evaluation Lab, and the a branch of the Makuhari IPv6 System Testing and Evaluation Center. These 3 centers are researching the fields of IPv6 network management, testing and evaluation methods for the interconnectivity of IPv6 networks, traffic dispersion control technology, backbone architecture and applied technologies, and related topics.

(4) R&D of Network Architecture Using Active Network Technologies

Research in this area is being performed by the Kita-Kyushu Gigabit Laboratory, Osaka University, and the Communications Research Laboratory. These 3 centers are active in research on efficient bandwidth reservation linking up multiple managed domains, dynamic route selection which responds to changes in network structures, end-to-end communication quality, precise measurement and estimation of internal network status, high precision stable network clock synchronization up to nanosecond accuracy, and general evaluation and testing of over the gigabit network.

4.2 Gigabit Laboratories

Gigabit Laboratories are located in Tsukuba, Keihanna, Kyoto, Okayama, and Kita-Kyushu. Users of these laboratories have to pay only minimal facilities fees for free access to mainframe computers and other equipment to conduct research on very high-speed networks and high performance applications, optical communications using integrated optical networks, and other areas.

4.3 Publicly Commissioned Research

As of the end of July, 2002, forty seven projects had been initiated through the public commission system. These are centered on telemedicine and education applications, but include a diverse array of interests which reflect the variety of the researchers themselves.

4.4 General Use Research

As of the end of July, 2002, one hundred thirteen general use research projects have gotten underway. Since use of the gigabit network requires agreeing to conditions outlined in the Shared Usage Agreement, proposals from single research institutions are rare cases. Most such proposals come from a group of multiple research institutions which form alliances to pursue single projects. Also, large scale R&D projects such as the Local IX Interconnectivity Project and the JB Project are carried out under the general use research framework.

4.5 Event Use

Event use occurs when researchers working on the JGN utilize the network for a relatively short period of time to perform demonstrations. Up until the end of July, 2002, this had occurred one hundred times. Since most general use researchers also make use of the network for such events, and one hundred thirteen general use researchers have performed such events over the network one hundred times, we can see that event use is almost universal among general use researchers, and that it forms a crucial aspect of the gigabit network. Some examples of event use are described below.

(1) Ryuichi Sakamoto's Opera "LIFE". In August and September of 1999, composer Ryuichi Sakamoto used the gigabit network to connect to concert halls in Frankfurt and New York in a world-wide performance.

(2) Information Processing Society of Japan's 61st National Conference Demo Session. At the IPSJ's (Information Processing Society of Japan) conference at Ehime University in October, 2001, high quality real-time video was bi-directionally transmitted between Kagawa Medical University

and Ehime University. This demonstration focused on the effectiveness of the policy control router installed at Ehime University.

(3) CRL's Live World Cup Soccer Broadcast. Taking advantage of the opportunity of the World Cup Soccer 2002 that was held jointly in Korea and Japan in April and June of 2002, scenes from the Incheon Stadium were piped through Korea's fiber-optic network, satellite relays, and the JGN to Kita-Kyushu city and Tokyo, where it was displayed on three panoramic high vision monitors in an ultra high resolution video transmission.

4.6 Overall Trends in Research and Development

When we look at all the research currently being conducted on the JGN, we see that it may be divided up into three main categories: network infrastructure, middleware technologies, and applications research. One trend visible is that the share of application-based R&D is slightly higher than that of infrastructure research.

5. JGN's Acceptable User Policy (AUP)

In order to insure that the JGN remains easy to use as a trial infrastructure for next generation Internet R&D, an AUP (Acceptable User Policy) has been crafted to ensure its smooth operation and that it does not give rise to problems in other connected networks.

1. Only Research and Development Use Permitted. In order to ensure that the JGN is used only for research, users are required to submit a research plan and agree to the Shared Usage Agreement before accessing the network. At the same time, research in which corporations use the network as they would the Internet in order to build up actual usage experience from which to make evaluations for improvement is also considered important.

2. Communication Quality is Not Guaranteed. Because the JGN is a trial infrastructure for R&D and not operated for real-world use, congestion experiments are performed daily, and changes in network settings for research purposes are always a possibility. Thus, it is impossible to guarantee communication quality to researchers who use the network.

6. Gigabit Network Symposium

A symposium is held once a year for researchers to report their results and interact with each other. From 1999 to 2002, these events were held in

Tokyo, Kita-Kyushu, Okinawa, and Sapporo. Scenes from these symposiums were transmitted over the JGN in DV (digital video) format to each Gigabit Laboratory (Shared Use Research Facility) as well as over the Internet in a form viewable to anyone with commonly available software. The symposiums are held jointly with related study groups and workshops from organizations like the IPSJ or IEICE, where there is much debate and discussion on issues related to the next generation Internet.

7. Summary

JGN is Japan's first very high-speed trial infrastructure for R&D. It is open to a wide variety of uses, including general use research, research controlled directly by the TAO, and publicly commissioned research, all of which yield remarkably varied research results and have allowed many people to interact with and contribute to the JGN effort. The JGN itself is also continually evolving as information and communication technology advances, which in turn spurs R&D efforts on to greater heights. The fruits of all this activity will be reflected in society in the near future. Experiments on international connectivity are also being performed through research alliances, allowing the JGN to contribute to international development of the next generation Internet in a concrete manner.

Reference

[1] JGN URL: http://www.JGN.tao.co.jp/index.html (as of 9/30/02)

Gigabit Network
T. Saito and H. Esaki (Eds.)
Ohmsha/IOS Press, 2003

Chapter 2

Directly Controlled TAO Research Using the JGN (TAO Research Centers at JGN's Birth)

Tomonori Aoyama[a], Kazunori Shimamura[b]
and Norio Shiratori[c]
[a] The University of Tokyo;
[b] Kochi University of Technology; [c] Tohoku University

1. The Goals of Directly Controlled Research

As was described in the previous chapter, TAO Research Centers were established at Makuhari and Kochi to conduct research using the JGN. Offices for research were also set up at the University of Tokyo and Tohoku University. The research at University of Tokyo office is conducted collaborating with the Makuhari Research Center, so here we will introduce the research activities of the Makuhari, Kochi Research Centers, and Tohoku University. As of July, 2002, researchers at these centers involved in TAO research included 1 director, 1 managing director, 2 sub-directors, 6 researchers, and 1 appointed researcher, for a total of 11 researchers. Fourteen research fellows are also assigned to provide support and advice as necessary.

In order to avoid overlap with publicly commissioned research, general use research, and event use, directly controlled research focuses its efforts on long-term network and application research, allowing it to focus exclusively on networking technologies for efficient use of gigabit class broadband networks and the applications which can take advantage of them. When the JGN began operation in April, 1999, it was an ATM based network, after which IPv4 connections came on line. Finally, in April, 2002, IPv6 connectivity was implemented. Therefore, directly controlled research has up until present been performed with ATM and IPv4 technology. IPv6 research will be discussed in the next chapter.

2. Research at the Makuhari Gigabit Research Center

2.1 QoS Improvement Research (Improving TCP Performance)
Research is currently underway on high performance TCP flow control

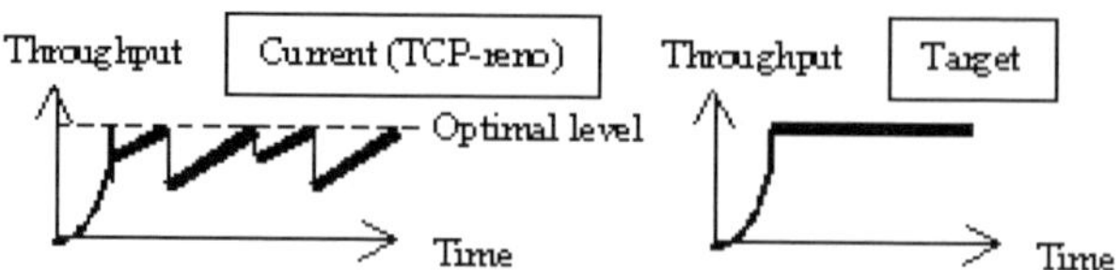

Fig. 1. Current and target throughput.

mechanisms. This includes both end to end QoS measurement technologies and TCP performance improvement efforts. It is said that TCP's performance declines as the delay and bandwidth product becomes larger [1] and/or when packet loss increases. Therefore, it is currently thought that TCP is not now able to take full advantage of available bandwidth even if packet transmissions are possible over long distances at high speeds of several hundred Mbps. Research is therefore underway to enable TCP to utilize the total bandwidth capacity available (Figure 1). It is known that the TCP Vegas is capable of high performance when it is used by all TCP flows, but that its performance declines when in competition with existing TCP flows.

The following two major factors are to avoid achieving a high TCP throughput with long-hole large bandwidth packet transmission.

1. According to the behavior of TCP's (transmission) window control mechanism, high performance packet transmission cannot be achieved without increasing the maximum window size to the product of bandwidth and approximate roundtrip delay times. In order to accomplish 460Mbps transmission between Kochi and Makuhari (roundtrip delay 26msec), we need a window of about 1.5Mbytes, which is about 100 times as large as the 8–32 Kbyte size commonly used in ordinary computers today.

2. Even if we can use the optimal transmission window size, the router's buffer could be filled up so as to cause packet losses. It is well-known that, even with only a few percent packet loss, the TCP's end-to-end throughput will suffer serious degradation.

In order to resolve these issues, a flow control mechanism that measures throughput and QoS so as to optimize the transmission window size using the packet pair mechanism is being developed. The system being developed, which uses TCP Vegas as a flow control mechanism, will revert to the existing TCP flow control mechanism when it detects resource competition.

Transmission experiments between Kochi and Makuhari over the JGN at 600Mbps PVC are underway. Equipment currently being used is as follows:

1. (PC) Router: A server PC running Windows NT 4.0 WS equipped with Giga Ether NIC and OC12 ATM NIC.

2. Server and client hosts: PC running FreeBSD 4.6.1 equipped with Giga Ether NIC. These computers have been adjusted so that maximum window size can be up to 4Mbytes.

Memory to memory throughput within a PC host was measured at 460Mbps, but packet loss occurred during packet transmission between Kochi and Makuhari with just increasing the TCP window size of server and client PCs. Further detailed investigation and the following research topics are being pursued:

1. Transmission of smoother packet flow, so as to reduce packet loss observed in packet transmission between Kochi and Makuhari.
2. R&D on methodology for QoS evaluation of packet transmission and for reverting to the conventional flow control mechanism when resource competition is detected.
3. High-speed, long distance transmission of application data in coordination with research on digital museum and digital cinema applications.

2.2 Dynamic Network Control

IP networks sometime experience the traffic concentration on certain routes. In order to overcome this issue, research on traffic engineering using MPLS (Multi Protocol Label Switching) is being conducted to improve network transmission capabilities. We have succeeded in dynamic traffic engineering which autonomously disperses IP flow over multiple paths (LSP: Label Switched Path) depending on network traffic conditions to avoid congestion. Other research involves moving network control from concentrated servers to individual LSR's (Label Switching Router) and setting alternate paths in advance to avoid congestion, which has been also successful. This research, in which LSR's autonomously perform distributed control functions, allows for both operating cost reductions (because no operator intervenes) and efficient use of network resources (because alternate paths are searched and selected when congestion is detected).

In this system, each LSR periodically floods its traffic statistics so that all ingress edge LSR's can collect such data from all LSR's and stores them in their database. This allows for the traffic through each link along LSP's to be analyzed. The edge LSR determines whether traffic through some links along certain paths has exceeded a predetermined congestion value, and if so searches for an engineering route as an alternate path. For this, an improved version of Dijkstra's shortest path search algorithm is used to find an LSP route with sufficient available bandwidth. Next, the RSVP Tunneling Protocol (RSVP-TE), which is a signaling protocol, is used to explicitly establish the appropriate LSP's. The ingress edge LSR then divides the ingress IP flow between the established engineering route and the default route (first recognized LSP) in order to evenly (or adequately) distribute traffic over the network.

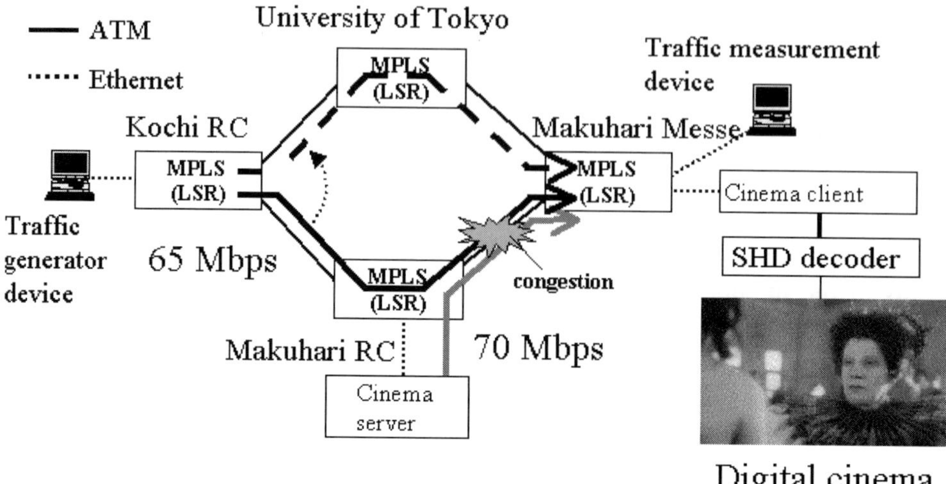

Fig. 2. Dynamic load distribution experiment system.

Since each LSR autonomously performs control functions on a distributed basis, the IP network can become over-controlled and traffic behavior can be destabilized. These behavior can be caused by the discrepancies between IP routing and the proposed LSP routing and be caused by a propagation delay (especially) in large scale networks. In order to prevent this phenomenon, the control parameters which yield a large effect on traffic behavior (statistical information communication interval, move granularity, and smoothing factor) need to be evaluated and the appropriate parameter values need to be selected for them.

MPLS routers with dynamic traffic engineering capability have been established at 4 locations on the JGN (i.e., Makuhari RC, Kochi RC, University of Tokyo IML, and Kita-Kyushu Gigabit Laboratory) in order to perform this evaluation for large scale networks. In order to establish a stable traffic control methodology, we define the following values used to measure this value and determine the optimum parameters for even traffic distribution; (1) *convergence time*, or time taken for even traffic distribution across the network, (2) control parameters such as move granularity (a measurement of traffic movement), and (3) smoothing factor (which prevents congestion from burst traffic and other dramatic changes of traffic patterns). As a result, we found that traffic distributed evenly across the network within about 300 seconds. Since rapid traffic fluctuations are rare in backbone networks, we considered this methodology to be sufficiently effective for load distribution. Even when super high definition (SHD) video systems (so-called digital cinema) were used on the network in accompanying application research, full functionality and throughput improvement were achieved when the parameters described above were applied (Figure 2).

2.3 Super High Definition Network Applications

In order to develop new applications using an SHD image transmission system, we are conducting research on digital museum and digital cinema applications using about 2000 scanning lines [2].

Digital archives using SHD images are being constructed by various research groups. Our digital museum is a system for research and education in the fields of Italian excavations and Edo "Kosode" kimonos. The purpose of this research is to elucidate the demands made on such a database and communication system, as well as the architecture of a prototype system. The digital data itself is created with a SHD camera on Brownie film. Through experiments on image throughput and resolution quality, we have proven that we can achieve a sufficient operational speed and quality for education and research purposes.

These images were then used in the construction of a remote database prototype system. The system finally developed allows users to easily search for images on a browser using thumbnails, keywords, or sketch images, the results of which are displayed on an SHD LCD monitor with a resolution of 2560×2048 pixels. The prototype of this image display system is shown in Figure 3, which shows the configurations of the database server (which performs searches and returns results) and the image server (which transmits image data), functions which have been separated to improve throughput. Because this system uses high resolution images, it is suitable for comparative research between geographically isolated institutions involving simultaneous display of multiple images and other high-speed network uses. Also, the sketch image search function in particular, which was developed for the Edo Kosode database, improves ease of use compared to traditional keyword searches. It has been proven to provide high search hit ratio on the distributed server and client system on the JGN.

Further, since one of the purposes of the SHD image system is the

Fig. 3. Digital Museum prototype.

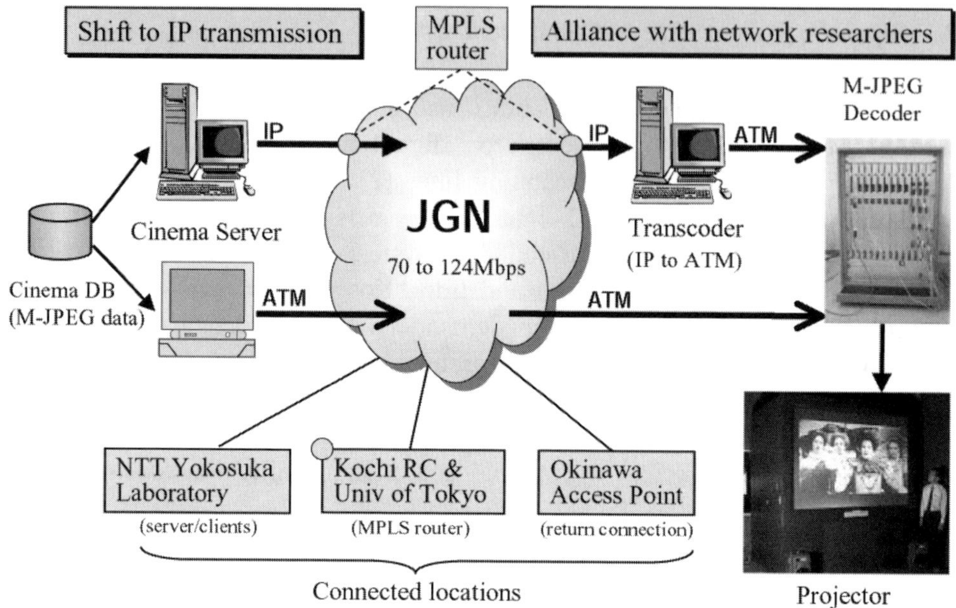

Fig. 4. Experimental digital cinema system.

digitalization of movies, which will require very high image quality, a progressive scan format with a spatial resolution of 2048×2048 with 24 bit pixels (8 bits each for R/G/B) and 60 frames per second is used. Transmission speed for raw images is several Gbps. These images are compressed with Motion-JPEG to the order of 100Mbps so they may be transmitted in real time. Movies on 35mm film, which have more than 20,000 frames, are digitized with a film scanner into a file of about 160Gbytes, which is then compressed to about 8–14Gbytes (70–120Mbps) with JPEG. A 100-inch high definition ILA projector is used for display.

As is shown in Figure 4, the experimental network used for this system connects the Makuhari Gigabit Research Center with the NTT laboratory collaborating in research (in Yokosuka) through the JGN Otemachi access point at line speeds of 124Mbps. The Okinawa access point was connected as well to study the effects of transmission delays (about 40msec), and it was proven that even such delays caused no dropped frames. The cinema server is capable of both ATM and IP transmission. The former was used in July of 2000 over iGrid and the later in October of 2001 over CEATEC JAPAN 2001 (Figure 2) in an experimental demo which transmitted data of higher quality than existing HDTV. In particular, the experiment on CEATEC included an MPLS router to alter transmission routes during congested times, resulting in the first time that cinema quality video had been transmitted smoothly with no disruptions. This experiment opened up new horizons in the realization of digital cinema broadcasting over large scale networks.

3. Research at the Kochi Communication Traffic Research Center

3.1 QoS and Line Control Technology

This research is aimed at improving QoS adapted to applications in very high-speed, large scale networks, and the diversification of new network media. The goal is development of technologies which can take advantage of networks increasingly capable of handling large amounts of traffic faster and faster.

Current research focuses on development of test models of devices for packet transmission suited to the JGN. Small packets processed frequently place burdens on the nodes, sometimes causing serious degradation of node port performance. In the prototype system, the edge nodes fulfills the function of aggregating multiple small packets generated on the access network into a single packet and transmitting them to the backbone network. This edge note role in the proposed system can relieve processing power required at the nodes on the backbone network (Figure 5).

However, with this packet aggregation technique at the edge node, some packets experience larger delay due to HoL (Head of Line) phenomenon. In order to resolve this issue, the "Priority Queuing with Assembly" is now being proposed as a way of reducing this influence by queuing time sensitive traffic first [3].

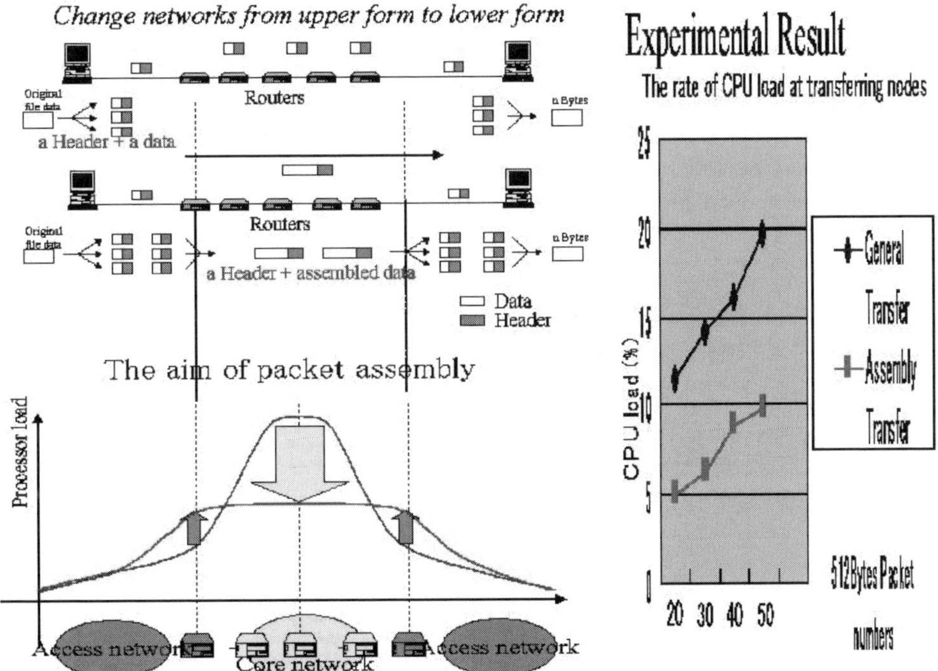

Fig. 5. Backbone node load reduction.

3.2 Very High-Speed Network Architecture Technology

At present, network transmission speeds are increasing at a rate that processor speeds cannot keep pace with. Also, the increasing clock speeds that have been the source of increasing processor speeds cannot continue their past trends because an increasing number of pipeline stages causes ever more processor loss and heat. Our research aims at utilizing processor resources more effectively. The elements of our methodology for the effective use of processor resources are as follows:

- A data-driven architecture prevents pipeline breaks which degrade processor usage efficiency.
- Improving processor capacity through multiple processors. Since there is a limit to the consecutive processing that can be made parallel at the processor level, this parallel processing will be described.
- Improving the load distribution algorithm on all processors when multiple processors are used in a single system architecture to improve system performance limits.

Data-driven processors (DDP) can be used to accomplish the first two elements. This mean that the research will require a dynamic load distribution methodology to equalize loads among processors (Figure 6).

Under previous research, multiple load prediction devices for dynamic load distribution have been developed and their effectiveness measured and evaluated under controlled circumstances. A example of today's practical applications using multiple processors on actual networks is a fast-response video encoder used in conjunction with packet assembly. Next, we will evaluate the effectiveness of dynamic load distribution (evenness of load and improved performance limits) mechanisms for the video encoder system in conjunction with packet assembly.

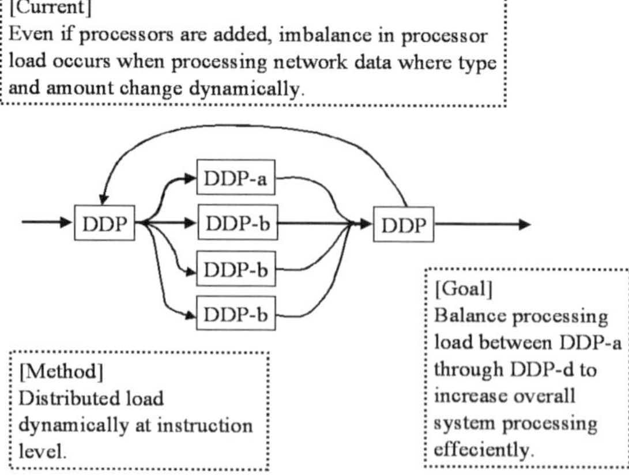

Fig. 6. DDP load distribution.

3.3 *Virtual Environment Communications*

We are conducting research on the 3D virtual reality environments, which will be used for browsing ordinary websites. To define a 3D space on the web, users must actively create links between virtual spaces, and it must be possible to move smoothly along links. The method we have researched enables users to create links to other (virtual) spaces at desired locations within their own space, view into the linked space without actually leaving their own space, and seamlessly move among spaces through links. (Figures 7 and 8.)

In the Virtual Space Link Method, individual links fundamentally work only for the user who created them (for viewing and moving between spaces), but they may also be shared among multiple users. This will allow multiple users to create and their own virtual space environments and combine them mutually in ways thcy desire to create a distributed shared

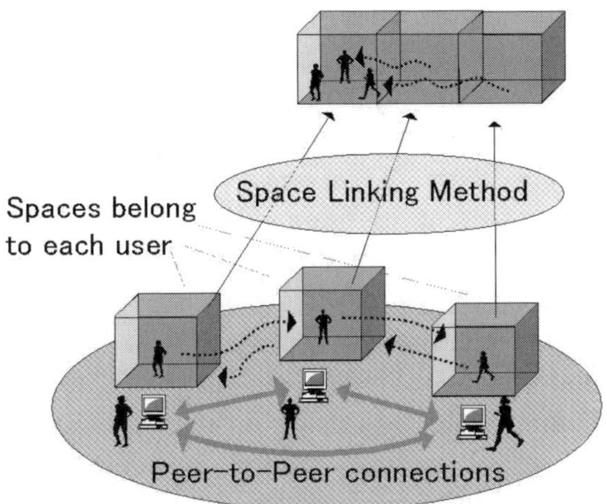

Fig. 7. Illustration of space linking method.

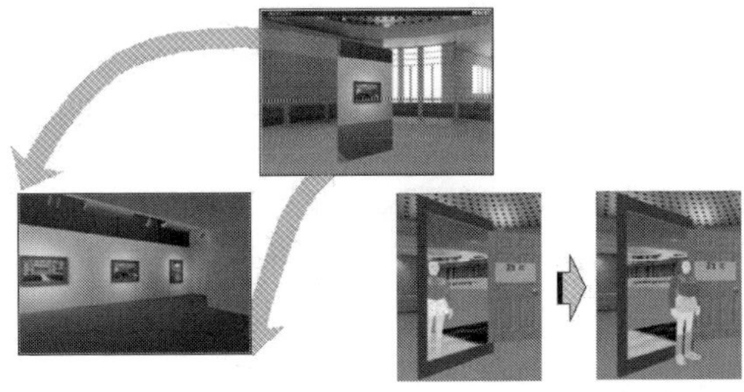

Fig. 8. Characteristics of space linking method.

virtual environment. Issues for future work include adding linkage rights management functionality and testing and evaluation in actual use.

3.4 Ultra-Large Distributed Database Technology

With networks which can handle larger volumes of data at higher speeds, video on demand is likely to become a reality. In a video on demand system, large amounts of varied video content would be formed into a database which could be streamed any time to user terminals. Recently, experimentation has been conducted in models which allow efficient transmission of video content by replicating content and storing it on geographically separate servers connected over the Internet, then transferring this content to the appropriate server when it has been requested by a user. This research will follow this model by tying video server groups located throughout the country together over the JGN, then create a television on demand architecture by transmitting and accumulating television broadcasts from different locations and distributing them on demand to users.

The television on demand system may be divided into 2 parts. The first part, or accumulation function, collects programs broadcast in different locations and transmits them to servers. The second part, or distribution function, streams this content to users when requested. It is necessary to transmit a large amount of content efficiently for the accumulation function to work properly. As is shown in Figure 9, we are considering a method whereby content is cycled through servers in a $S1 \rightarrow S2 \rightarrow S3 \rightarrow S1\cdots$ fashion. This method will allow multiple content files to be accumulated and transmitted in real time while only using up the bandwidth of a single file.

At present there are 2 methods being considered for the distribution function which will use replicated content on multiple servers to guarantee the same quality as the original even if congestion occurs during streaming

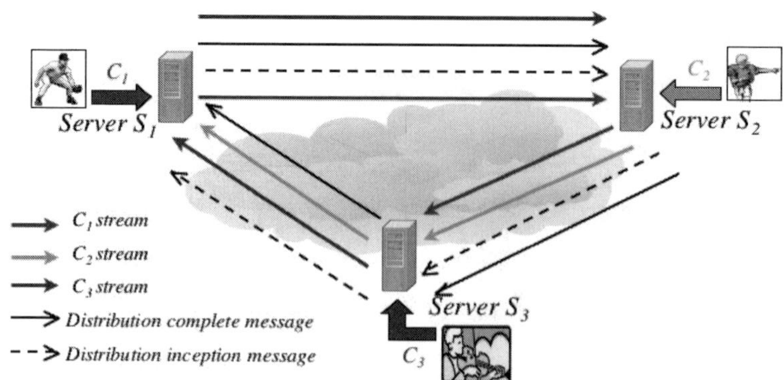

Fig. 9. Content cycling distribution method.

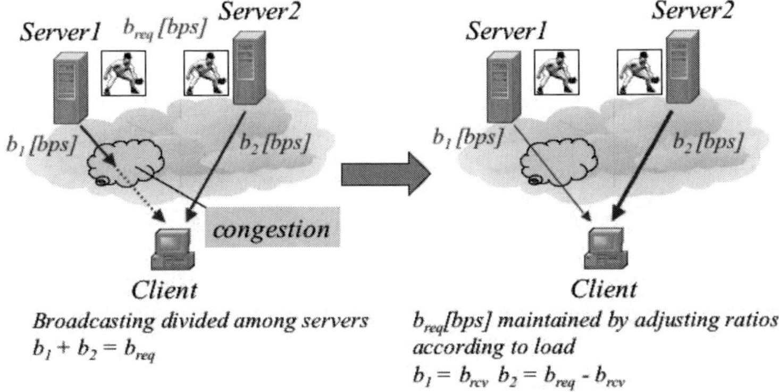

Fig. 10. Content distribution control to maintain quality.

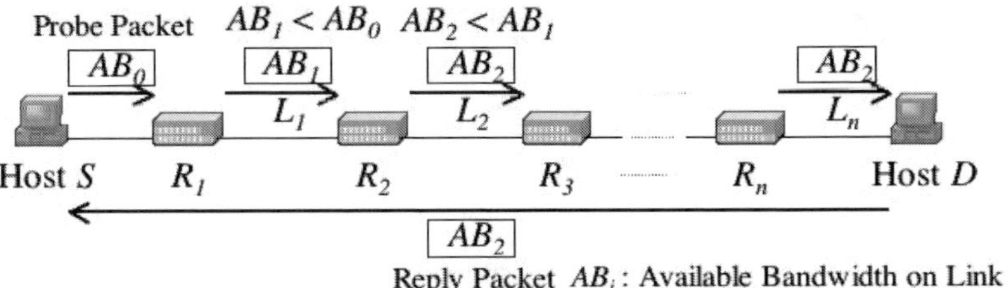

Fig. 11. Available bandwidth measurement method.

(Figure 10). These involve measuring the available bandwidth in the routes available and either switching to the route (server) with the greatest present capacity or adjusting the transmission levels from each server according to available bandwidth. In this case, methods for measuring available bandwidth become an issue. We have proposed and are designing the measurement method shown in Figure 11.

In the future, we are planning on testing the effectiveness of the above-mentioned content distribution methods, streaming methods, and available bandwidth measurement methods toward the goal of creating the architecture for a television on demand system [4].

4. Research at Tohoku University

4.1 Network Management and Control

Tohoku University is utilizing the JGN as an experimental infrastructure to pursue R&D in 3 subjects: 1) flexible QoS control in high-speed networks, 2) visualization systems for network information, and 3) storage systems for high-speed video servers.

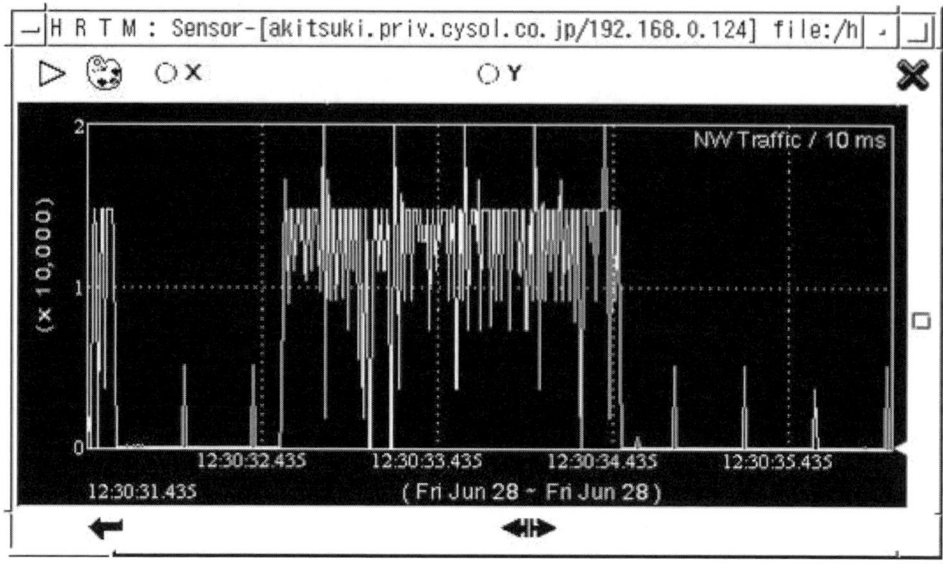

Fig. 12. High precision traffic information according to HRTM.

For subject 1, we conducted R&D into a Network Information Warehouse (NIWH), which will offer high quality network information to network applications [5]. We also developed the central function of the NIWH, a High Resolution Traffic Measurement (HRTM) system [6]. Figure 12 illustrates an example of real time remote precision traffic information output visualized with HRTM on the JaNI system, which is described below.

For subject 2, the results from subject 1 are used to develop the network information visualization system JaNI (JGN next generation Network Information System), which is presently being submitted to the IETF (Internet Engineering Task Force) for consideration as a standard.

For subject 3, we conducted research into the characterization of traffic as a fundamental technology for network traffic management. We also performed research on efficient congestion control technology which did not impede link usage efficiency and methods of distributing bandwidth fairly among multiple connections.

Here we would like to take the opportunity to introduce the JaNI network information visualization system in more detail.

4.2 JaNI Network Information Visualization System

The advent of the Internet has revolutionized the way we create, process, access, and distribute information. However, traditional Network Information Centers (NIC's) have fallen short in many respects. Even aside from

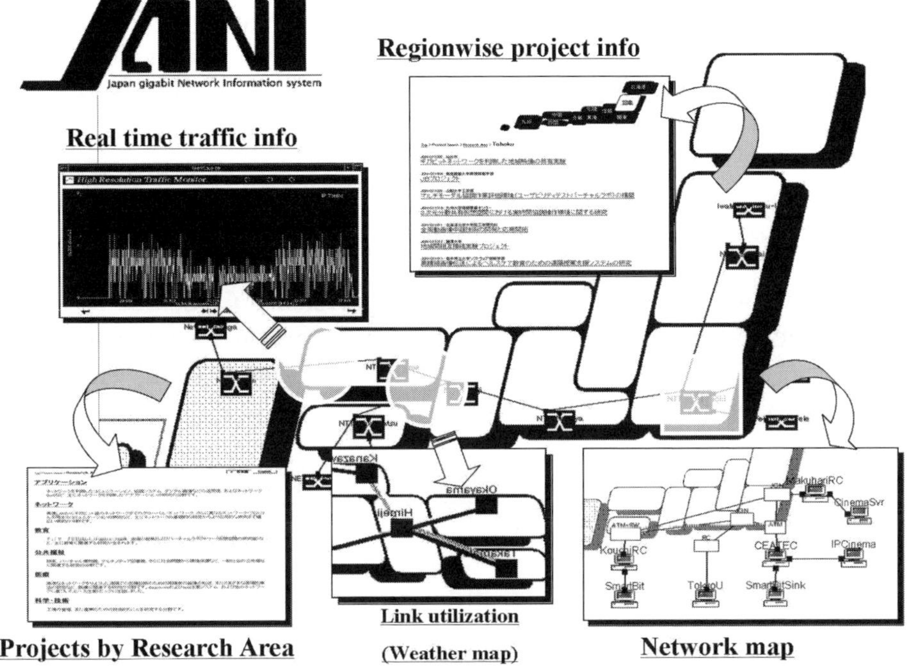

Fig. 13. JaNI system functions.

traditional NIC's directly related to profit-making activities, their policies and objectives have not always been clear. Nor have they been sufficiently useful to researchers or general users. In contrast, JaNI will offer a high level of functionality even to general users.

JaNI has enabled access to a wide variety of network information to users from specialists to general users. It provides the following kinds of information:

A. Highly time-precise traffic information (required installation of HRTM probe)

B. Traffic Information for remote locations

C. Network maps by project

D. Project distributions by region and research field

Figure 13 shows a more complete list of functions.

Technically speaking the JaNI system employs the Internet's standard management protocol SNMP. JaNI has the 3 following characteristics:

1. On demand, real time information collection

2. Millisecond-level data collection from remote probes

3. Offers high level network information to users

In addition to this functionality, various efforts have been made to insure ease of use as well. For example, browsers are used as an interface so that information can be accessed from anywhere. Also, graphical network maps are used as the basis for information access, giving an intuitive feel to program operation. Finally, because some users may have malicious intent behind there use of such information, JaNI employs access controls based on SNMP technology.

4.3 Comparison of JaNI with Other Systems

Table 1 is a comparison of JaNI with existing systems. Compared to existing systems, the JaNI system enables extremely powerful functionality. This includes the ability to collect network information remotely, on demand, and in real time with millisecond-level precision.

Table 1. Comparison of JaNI with other systems

	Mode of Service		Online User Interface	On Demand Info Collection	Resolution (Hi — med — low)	Information other than traffic
	Remote+ Real time	Archived Info				
CAIDA CoralReef	X	O	X	X	● Hi (milli secs)	O
MAWI Repository	X	O	X	X	● Hi (milli secs)	X
MRTG	△ 5 mins	△ (Aggr)	△	X	● med (30 secs)	X
Champion	X (next day)	O	△	X	● low (10 mins)	X
JaNI	O	O	O	O	● Hi (milli secs)	O

5. Future Plans

Three and a half years have passed since research controlled directly by the TAO began on the JGN, with another year and a half remaining. Up until now all research has been conducted separately, but in the remaining time the various projects will be integrated as each nears completion. This process will be accompanied by experimentation and testing of broadband service measurement and control, and the improvement in quality and functionality of resulting applications.

Acknowledgements

The research described in this article was performed by the Tadashi Enomoto, Junji Suzuki (presently at NTT Network Innovation Laboratories), and Takao Ogura of the Makuhari Gigabit Research Center; Hiroshi Kato, Kisho Takamatsu, Toshikatsu Kanda, and Takuji Nakahira at the Kochi Communication Traffic Research Center; and Goutam Chakraborty at Tohoku University. All of these research kindly took time away from their busy schedules to aid in the writing of this article, for which we are deeply appreciative.

References

[1] Kleinrock, L.: The Latency/Bandwidth Tradeoff in Gigabit Networks, IEEE Communication Magazine (April, 1992)

[2] Suzuki, J., Kato, H., Takamatsu, K., Shimamura, K., Shiratori, T., Aoyama, T. and Saito, T.: Application Research Under Control of TAO, Journal of the Institute of Image Electronics Engineers of Japan, Vol. 30, No. 6, pp. 747–752 (Nov. 2001).

[3] Kanda, T. and Shimamura, K.: Load Balancing Technique for Node Processors by Packet Assembly, 2001 Asia-Pacific Symposium on Information and Telecommunication Technologies (APSITT).

[4] Nakahira, H. and Shimamura, K.: End-to-End Bandwidth Availability Measurement Method Using ICMP, FIT (Forum of Information Technology), 2002 Information Technology Letters, Vol. 1, pp. 207–208 (September, 2002).

[5] Saito, T., Mansfield, G. and Shiratori, N.: Network Monitoring in the Large: Distribution and Integration, International Journal of Computer and Information Science, Networking and Parallel/Distributed Computing, Special issue of SNPD'01, Vol. 3, No. 3 (2002).

[6] Mansfield, G., Saito, T. and Shiratori, N.: A Technique for SNMP Based High Resolution Remote Monitoring, International Journal of Network Management (to appear).

Chapter 3

The Japan IPv6 Network

Hiroshi Esaki[a], Satoshi Katsuno[b],
Kazumasa Kobayashi[c] and Yukiji Mikamo[b]
[a] The University of Tokyo;
[b] Telecommunications Advancement Organization of Japan;
[c] Kurashiki University of Science and the Arts

Abstract. Research on IP version 6 (IPv6) has been spurred on the growing shortage of IPv6 addresses and the explosion of network routes.

The TAO (Telecommunications Advancement Organization) has used multi-vender commercial routers to implement IPv6 on the JGN (Japan Gigabit Network) an experimental network for the development of the next generation Internet, resulting in a unique IPv6 network. Twenty eight nodes have been established at points throughout the country with routers to provide IPv6 access at a total of forty seven locations so that tests and experiments may be performed to hasten the transition to and development of products for IPv6.

This chapter will give an overview of the JGN IPv6 network, its current operation, research and development performed on it, as well as a snapshot of its experimental uses.

1. About the JGN IPv6 Network

In order to develop and deploy next generation very high-speed networks, the JGN [1] was constructed as an R&D platform for network operations and architecture as well as high performance applications. The network is open to universities, research organizations, governmental organizations, local government, and corporations. IPv6 routers have been installed on the JGN to enable R&D of IPv6 technology, the core Internet protocol for the next generation Internet. On October 1, 2001 the JGN IPv6 network came on line.

IPv6 will support the security and reliability required of an IT infrastructure, as well as enough address space to preserve the end to end model on which the Internet is based and developed. Therefore, IPv6 can resolve the recent address shortage and route explosion issues which have arisen with IPv4. The standardization of IPv6, which will become the basis of emerging Internet architecture, was completed several years ago in the form of RFC 2460 and other specifications [2]. At present, network architecture and other operational functions are in the process of being standardized, and commercial routers from different companies are being deployed.

The JGN IPv6 network was established as an R&D platform for network operation and architecture as well as high performance application technology. Filling out the necessary usage forms [3] allows researchers to use the network to perform R&D. Further, the Okayama Interoperability and Evaluation Lab and its partner Makuhari have been established in order to perform interconnectivity testing and evaluation in parallel with the construction of the network, while the TAO Tokyo Research and Operation Center has been established in Tokyo to develop operation and management technologies for IPv6 network equipment.

1.1 Network Structure

The JGN IPv6 network is a purely native IPv6 network which did not include any IPv4-only elements, but it can also utilize dual stacks (including both IPv4 and IPv6). Since the JGN is based on ATM technology, the JGN IPv6 network fundamentally overlays the ATM network.

The JGN IPv6 network consists of 57 sites throughout Japan, 28 of which are router sites and 29 of which are bridge sites. The network topology is shown in Figure 1.

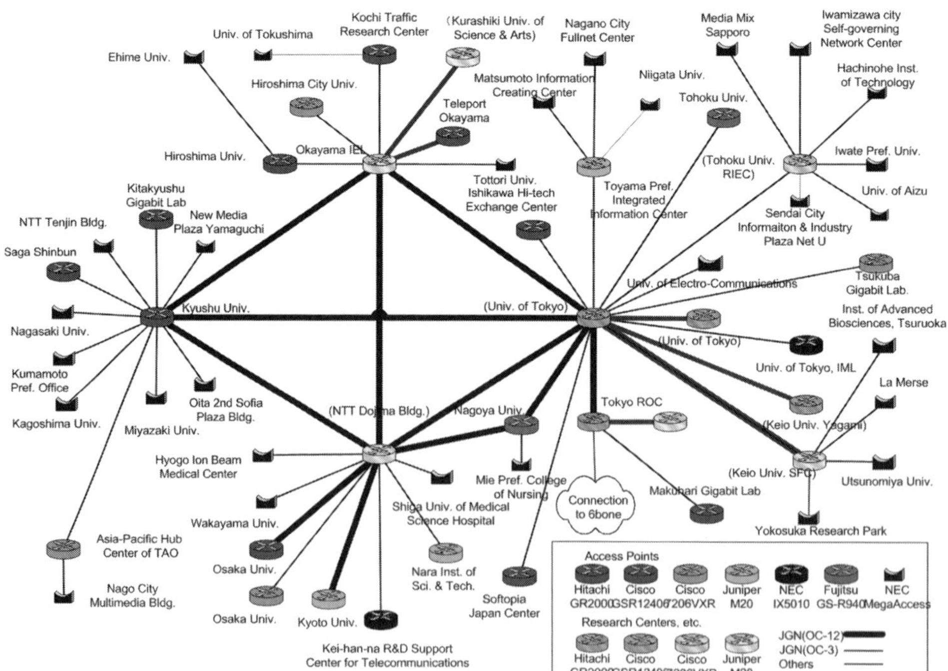

Fig. 1. Connection structure of JGN IPv6 network.

Of the router sites, the University of Tokyo, Teleport Okayama (built alongside the Okayama Interoperability and Evaluation Lab), Dojima, and Kyushu University have been functioning as the four core sites. Each core site is connected to each other through JGN ATM (OC-12) link, forming the backbone of the JGN IPv6 network. Fundamentally, all other router sites are connected to one of these four, and each bridge site is connected to a router site.

1.2 Multi-Vender Environment

Routers from multiple manufacturers have been set up at each site (partially depending on the technical requirements at each individual sites) to test and evaluate interconnectivity using actual network environment. In this respect, the operation of the JGN IPv6 network itself has become a large scale interconnectivity experiment utilizing a diverse array of commercial routers. Names of these routers are listed in Table 1.

Table 1. Routers in operation on the JGN IPv6 network

Vender	Model	Quantity
Cisco Systems	GSR12406	3
Juniper Networks	M20	8
Hitachi	GR2000-gH	13
Fujitsu	Geostream R-940	3
NEC	IX5010	3

Because these routers are provided by different manufacturers, they differ in details of IPv6 compliance and implementation, reflecting different viewpoints about the next generation Internet market when each vender began their development. Therefore, functionality occasionally differs across all the routers. Software upgrades for each router both to patch bugs and to provide new functionality are performed as soon as they are made available.

1.3 Network Characteristics

The JGN IPv6 network leverages the special characteristics of ATM networks, meaning that ATM paths are configured between each node to define a logical network architecture. It is possible to manipulate network architecture freely by changing these network paths (i.e., ATM PVC) for high performance routing or other experiments. Current network topology is a result of designs intended to establish the JGN IPv6 network and enable its operation as soon as possible, but practical operational experience will increase its level of topological and functional complexity.

Also, in preparation for technical difficulties arising during operation, every router site has IPv6 compliant ATM-Ethernet bridges in order to perform the same functions as bridge connection sites. If the upstream router nodes work incorrectly, the ATM-Ethernet bridges can be used to accommodate and forward the traffic from adjacent router sites.

1.4 Address Space and Routing

NLA addresses (3ffe:516::/32) in pTLA address blocks from the WIDE project are used for the JGN IPv6 network's address blocks. Methods for allocating addresses to router and bridge sites are one research subject evaluated on the network. Though JGN IPv6 network uses the WIDE NLA, the JGN IPv6 network owns its own AS (autonomous system) number (AS17934) so that experiments on routing path information exchange with other IPv6 networks can be performed independently. When the JGN IPv6 network came on line, it used a combination of RIPng and static routing, but broadband routing experiments on dynamic routing with IS-IS and OSPF are also being planned while taking router characteristics into consideration.

1.5 External Connectivities and Dual Stack Operations

A connection diagram for the JGN IPv6 network is shown in Figure 2. External connectivities are maintained both through the IPv6 router at Tokyo Otemachi with WIDE NSPIXP-6 [4] and with the worldwide experimental IPv6 network 6Bone. Other external connectivities are also maintained with local IX's such as NSPIXP-3 (Osaka Dojima) and OKIX (Okayama).

The JGN IPv6 network offers IPv6 connection services, with IPv6/v4 conversion servers developed by the KAME [5] project running atop IPv6/v4 dual stack segments established at external connection nodes, allowing external connections to hosts on the JGN IPv6 network from IPv4 networks. Also, DNS servers, WWW servers, and mail servers using IPv6 stacks developed by the KAME project are in service and provide information to managers of each router node.

The connectivity between the USA's next generation Internet network Abilene, which is also a native IPv6 network, and the JGN IPv6 network were established as Abilene came on line.

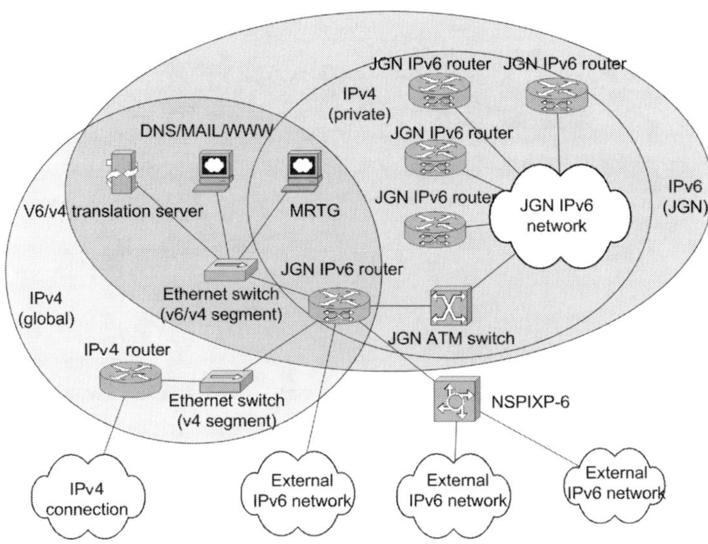

Fig. 2. Connection topology for JGN IPv6 network.

1.6 Access Points

Of the 57 access points on the JGN IPv6 network, the 47 which are not in facilities directly controlled by the TAO are all open-use access points which offer IPv6 connectivity through Ethernet (10/100Mbps) to users. The IPv6 segments used for these connectivities belong to the same Ethernet segment used to serve users of the router sites upstream of each bridge site.

Figure 3 shows a network topology diagram for the router sites, the access points for which consist of an IPv6 router and an Ethernet switch to provide IPv6 connectivity to end users.

Users willing to directly connect to the router node can use the access points directly connected to the Ethernet switches 10/100BASE-T ports at the router site. Each router site maintains /48 address for end user use.

Figure 4 shows a network topology diagram for the bridge sites. Bridge sites utilize ATM-Ethernet bridges which can transmit/bridge IPv6 packets,

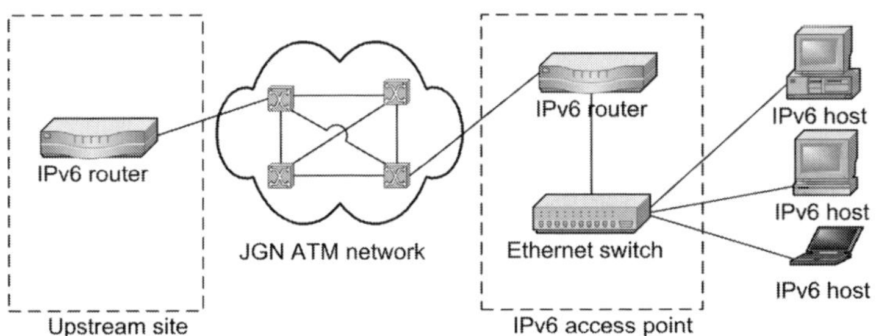

Fig. 3. Router site structure.

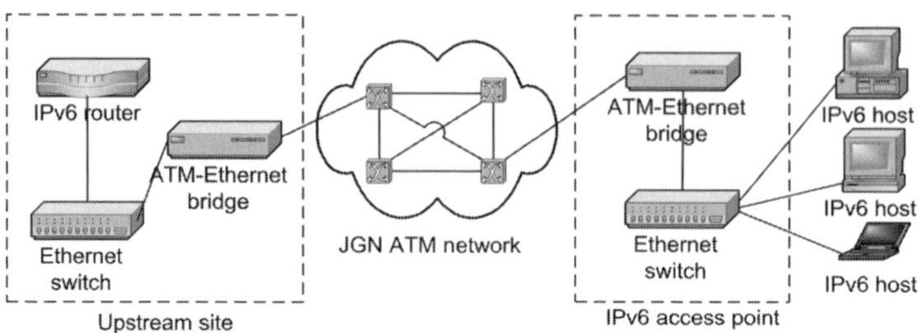

Fig. 4. Bridge site structure.

offering an environment where IPv6 can be used by extending the IPv6 segment from one of the router nodes. There are plans to delegate the operational responsibilities of the IPv6 network (bridge site) to the network manager of the router site connected to the ATM bridge. This delegation will include tasks such as address management and assigning default gateways, and will be able to be performed as well as for the service segments of the manager's own network.

On the JGN IPv6 network, network administration and monitoring activities are also a part of the experimental efforts. Therefore, new router nodes have also been established in locations where the TAO had not set up JGN IPv4 network access points. This provides many opportunities for mastering of daily technical skills required to run an IPv6 network that may not necessarily apply to IPv4 networks. There are a wide variety of network configurations on the Internet at large which developers could not configure at their laboratory nor imagine during development. Therefore, for commercial network operators and for network equipment manufacturers, obtaining feedback directly from network managers on the JGN IPv6 network is a valuable source of information that can help prevent and solve potential technical problems.

2. Research and Development on the JGN IPv6 network

2.1 Testing IPv6 Router Interconnectivity

Not all of the IPv6 routers currently in use on the JGN IPv6 network have good interoperability. Unlike the IPv4 network, there has yet to emerge a manufacturer of network equipment which sets the *de facto* standard for the industry. Therefore, the current state of interoperability for IPv6 equipment greatly resembles that of IPv4 equipment 10 years ago.

Therefore, the Okayama Interoperability and Evaluation Lab has been established to test the interoperability of the IPv6 routers and switches on the JGN IPv6 network with cooperation from various venders and research organizations. This center tests and evaluates all of the routers which are in use in various places on the JGN IPv6 network, and performs laboratory tests to validate interoperability before introducing them on the active JGN IPv6 network

New versions of software for IPv6 routers are released frequently, with very close communication between the manufacturers and JGN IPv6 staff. This is because of efforts both to introduce new IPv6 functionality and to fix bugs in the software on a short cycle. Wherever possible, feedback is given directly to the manufacturer in order for the evaluation and testing conducted on the JGN IPv6 network to be incorporated in products on the market.

2.2 IPv6 Network Management Technologies

The JGN IPv6 network is an experimental network for testing routers from multiple manufacturers, so the IPv6 system does not have sufficient standardization and implementation of network management and security measures for operation in a purely IPv6 environment. Since the IPv6 network management protocol SNMPv3 [5] is still in the process of standardization, the IPv6 based network management features of routers from different venders sometimes differ from each other. Similarly, network management applications running over IPv4 networks have not yet been ported to the IPv6 environment. Since functionality such as the collection of IPv6 address traffic information and 128bit address formats have not yet been enabled, many practical difficulties regarding network operation remain for the moment. Thus, the tools and software for such network management functions have had to be developed in-house to operate the JGN IPv6 network.

The TAO Tokyo Research and Operation Center is responsible for developing software necessary for network management as well as actual operation of the JGN IPv6 network. They are also developing network management tools and traffic information monitoring systems.

2.3 JaNI, a Tool for Network Information Management

Deployment of the IPv6 network's management information base (MIB) is generally running behind schedule because data transmission has received priority over management functions. There is one management agent, RMON, which allows locations monitored to be set freely, but filter and other settings for this software are difficult to perform, and no version exists yet for IPv6 networks (both transmission and information collection functions). Therefore, it cannot yet be used for IPv6 network management.

To meet with this lack of functionality, JaNI Gigabit Network Information System), an IPv6 network information management tool, and a new multi-protocol agent for traffic information collection are being developed. This agent will passively monitor network traffic and provide the information to management applications. We applied the SNMP protocol to this function, so a CpMonitor MIB is under development.

A test system for this agent, which can accommodate both IPv6/v4 data transmissions, was developed at Tohoku University TAO Gigabit Network Research Laboratory in order to perform realistic and practical experiments. Using this test system, traffic information for TCP, UDP and ICMP both over IPv6 and IPv4, as well as the same information for different applications, can be monitored.

3. Examples of Practical Experiments using the JGN IPv6 Network

Various practical experiments are performed on the JGN IPv6 network as part of the process of transforming it into a stable and working network platform. Here we would like to introduce two examples of these experiments. One is Miyagi IT Forum IPv6 multicasting and the other is high quality, real time teleconferencing at the Shikoku Four Prefecture United e-Meeting.

3.1 IPv6 Multicasting at the Miyagi IT Forum

At the Miyagi IT Forum held in May, 2002, the TAO and the JGN IPv6 network operation team performed a multicasting experiment using commercial IPv6 routers. The Miyagi IT Forum is sponsored by Miyagi Prefectural government and other local organizations. Its goal is promoting IPv6 technology and accelerating activities in broadband networking in regional networks. High quality still and video image broadcasting experiments mixing both unicasting and multicasting over the JGN IPv6 network were performed at this forum. This demonstration was the first multicasting experiment utilizing such a large scale practical IPv6 network ever.

Not all of the routers on the JGN IPv6 network possessed multicasting functionality at the time, so a temporary network topology was configured for the experiment. The network topology is shown in Figure 5. The Okayama Interoperability and Evaluation Lab worked as the core site, with four multicasting sites branching off from it. The rest of sites were connected

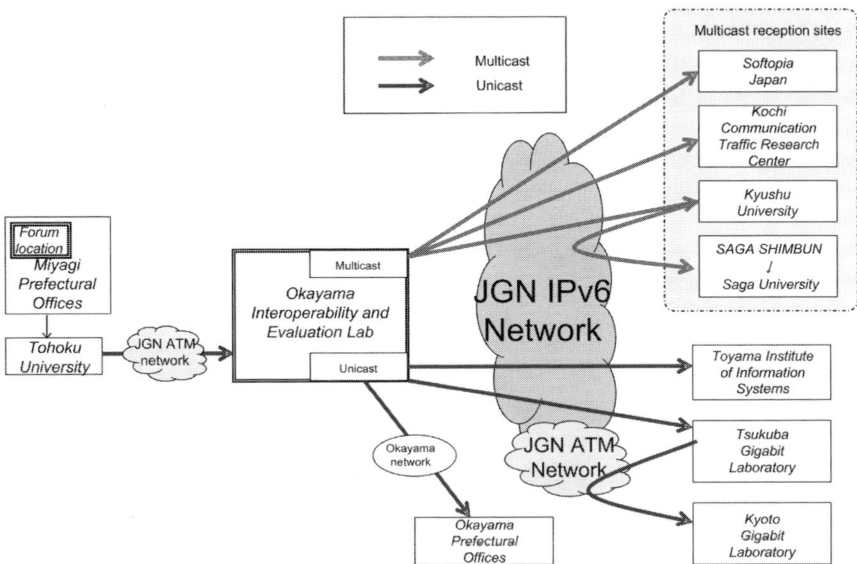

Fig. 5. IPv6 multicasting experiment at the Miyagi IT Forum.

through unicast connectivity. High quality images were transmitted from the forum location at the Miyagi prefectural offices to the Okayama Lab, from where they were broadcast to all the other locations.

3.2 Teleconferencing at the Shikoku Four Prefecture United e-Meeting

The TAO and the JGN IPv6 network operation team cooperated to create a 4-way teleconferencing system over the JGN IPv6 network at the Shikoku Four Prefecture United e-Meeting held in June, 2002. The Shikoku Four Prefecture United e-Meeting is held jointly by the Shikoku Communications Division of the Ministry of Public Management, Home Affairs, Posts and Telecommunications and the Gigabit Network Shikoku Conference. It enables telecommunication conferences among IT related personnel from Shikoku's four prefectures (Kagawa, Tokushima, Ehime, and Kochi) and university researchers. JGN access points have been established in each of Shikoku's prefectures (Shikoku 1–4), of which Shikoku 3 and Shikoku 4 offer JGN IPv6 service, meaning that some prefectures do not offer IPv6 access. Thus, this event required expansion of the JGN IPv6 network to encompass each access point and the universities which served as the meeting locations. Kagawa University, the University of Tokushima, Ehime University, and Kochi University of Technology were selected.

To provide IPv6 connectivity, ATM paths were first established from the ATM switch at Shikoku 1 to access points Shikoku 2–4 and the TAO's

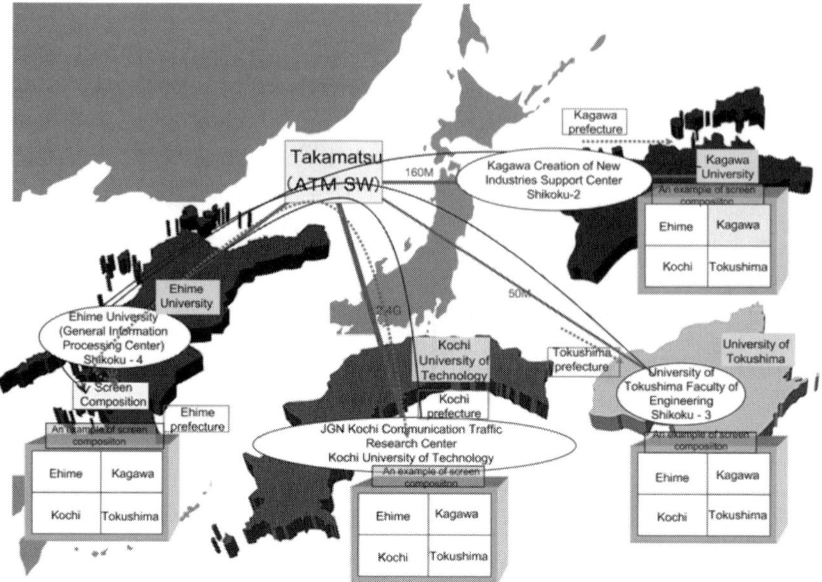

Fig. 6. Shikoku four-prefecture United e-Meeting over JGN IPv6.

Kochi Communication Traffic Research Center. Next, IPv6 address spaces were allocated from the IPv6 router at the Kochi Center. The Digital Video Transmission System (DVTS) was used as the image data format to transmit the high quality video images from Ehime University, connected to Shikoku 4 to the other three universities. Image synthesis and re-broadcasting was also performed at Ehime University. This system achieved 4-way television-quality teleconferencing at the four meeting locations in four prefectures (Figure 6).

4. Future Goals

The JGN IPv6 network is now accessible through 47 nodes to enable a wide variety of R&D activities on it. By constructing its experimental IPv6 network environment stretching over the entire country, it is hoped that the JGN IPv6 network will contribute to the early development and deployment of the next generation Internet, the creation of new industries using very high speed network technologies, and the generation of new employment opportunities.

In order to complete the introduction and deployment of IPv6, everything currently possible on the IPv4 network must first be enabled on IPv6. Next, developments in the following areas are also necessary:

- Address allocation methods between IPv4 and IPv6
- Upgrade and migration methods for hosts and routers
- DNS adaptation and deployment for IPv6
- Migration plans for individual sites to IPv6

• Migration of the entire Internet to IPv6

The operation of the JGN IPv6 network itself provides opportunities to use a huge address space and to solve many technical issues with IPv4. A working network is indeed necessary for development, deployment, and validation of functions such as network management, security, load distribution, and video streaming. In this way, the JGN IPv6 network can be a practical and useful platform to develop a next generation broadband network which both reflects market needs and provides feedbacks to the market.

The JGN IPv6 network will contribute toward the development and deployment of IPv6 technologies by (1) actively interconnecting with other IPv6 networks, including overseas networks, (2) testing and evaluating network operation and management techniques, and (3) giving feedback to the industry about results of performance and interoperability tests on IPv6 equipment. The JGN IPv6 network also offers a valuable opportunity for operator to acquire IPv6 network management techniques and know-how through first-hand experience of operating the network itself.

Acknowledgements

We would like to express our gratitude to people at the Ministry of Public Management, Home Affairs, Posts and Telecommunications, the TAO, and everyone else involved in making the JGN IPv6 network possible.

References

[1] TAO: Japan Gigabit Network, http://www.JGN.tao.go.jp/
[2] Deering, S. and Hinden, R.: Internet Protocol Version 6 (IPv6) Specification, IETF RFC2460 (December, 1998).
[3] TAO: On Beginning IPv6 Service Trials on the Research Gigabit Network, http://www.JGN.tao.go.jp/org_tec/ipv6_start.html (September, 2001).
[4] Wide Project: NSPIXP-6, IPv6-based Internet Exchange in Tokyo, http://www.wide.ad.jp/nspixp6/
[5] KAME Project: KAME Project, http://www.kame.network/ (December, 1997).
[6] Case, J., Mundy, R., Partain, D., and Stewart, B.: Introduction to Version 3 of the Internet-standard Network Management Framework, IETF RFC 2570 (April, 1999).

Chapter 4

Regional Interconnection Experimentation Project

Eisuke Hayashi[a], Masahiro Hiji[b], Yutaka Kikuchi[c],
Ikuo Nakagawa[d] and Kazuhiro Yatsushiro[e]

[a] *Reitaku University;* [b] *Tohoku Internet Association;*
[c] *Kochi University of Technology;* [d] *Intec NetCore Inc.;*
[e] *Yamanashi Women's Junior College*

Abstract. Regional interconnection experiments conducted by the Regional Internet BackBone (RIBB) project are research and development efforts for JGN (established and administrated by the TAO), whose goal is achieving a synergistic effect on regional Internet activities by connecting regional networks. This chapter will describe the RIBB and its activities, including the particularly active area of broadcasting video demonstrations from regional events nationally. Finally, the effect JGN has had on regional Internet activities through these efforts will be discussed.

1. Origins

According to the meeting's record of proceedings, the meeting was held at Reitaku University's Tokyo Research Center in Nishi-Shinjuku. It was May 1, 1999, and forty people involved in regional IX attended. After a heated debate that lasted more than 4 hours, the Regional Interconnection Experimentation Project was finally born.

Even before the advent of JGN, regional Internet activities existed in many regions, including Hokkaido, Miyagi, Yamanashi, Toyama, Nagoya, Okayama, Hiroshima, and other prefectures. Each of these has its own history, and has certainly contributed to its region in its own way. Further, these local groups are well known to each other, with information and experience exchanged at the Regional IX Operators' Meetings, which as been held seven times already, and the special interest group DSM of the IPSJ. However, the regional networks themselves had not been directly connected, and traffic and technical expertise had not been exchanged in an organic way.

Needless to say, everyone at this meeting was quite enthusiastic about the prospect of a high-speed national experimental network which they could use for free. Indeed, it seemed as though this network had been created especially for them. We decided on beginning the project without fully understanding what JGN was and whether or not we could even really use it.

1.1 Naming

The next step after birth is naming. This step is crucial to the success of any project, so the appointment of the chief naming officer was an important task for the members.

According to a presentation we had heard previously, there is a project concerning JGN called "JB". We were not able ask what "JB" stood for, but when we saw the project's logo, we figured out that "J" stands for a country in East Asia and "B" for a portion of the anatomy common to all vertebrates.

With other projects named like this, our naming had to be stylish as well. Our chief naming officer decided on RIBB (Regional Internet BackBone). As will be explained later, our network spreads out evenly from Tokyo, so a word that would imply a center would not fit. We desired an image that would go beyond such a center-based focus, such as the idea of something surrounding the center like ribs surrounding the heart, and that was what our naming officer settled on. Thus was the name RIBB born. We decided that all the domain names with only one "B" would likely be taken, so we kept the same pronunciation and added the extra "B".

1.2 Our Expectations of the Future

At the time, Mr. Nakagawa, who proposed the idea, explained the background to us this way: "Think five years into the future. Did commercial ISP's exist in Japan from the very beginning? Lines overseas were only 0.2–0.5Mbps. Now we have OC3's. So, imagine what things will be like 5 years in the future". Mr. Nakagawa already had his ideas of what things would look like 5 years in the future, and they are being borne out a mere 3 years later: high-speed and diverse networks, fragmentation of ISPs, advancement in routing technology, and IPv6 becoming pre-eminent (Table 1). Thus, one aspect of the background for our project was preparing for what we today call "broadband".

Table 1. Features of internet infrastructure

Date	ISP's	Overseas bandwidth	Backbone	Access Lines	IP Version
~1994	Initiative by Universities	0.2–0.5M	SDH	Dialup	v4
1999	Spread of commercial ISP's	155M	ATM	ISDN, CATV	v4
2002	ISP consolidation	655M–2.4G	PoS (OC192), GbE	ADSL, CATV	Mostly v4, some v6
2004 [a]	ISP specialization increases	10G	PoS (OC768), 10GbE	FTTH, ADSL, CATV	Mostly v6, some v4

[a] Predicted.

1.3 More Background

In order to fly from one regional area to another for a research meeting, you often have to connect through Tokyo's Haneda airport. Even if you are flying from, for example, Kochi to Ishikawa, you have to fly two sides of the triangle to get to your destination. If you look at the air route map on the front of the Japan Travel Bureau's timetable, you can see graphically how most of the travel routes converge at Haneda. This map does not show numbers of flights, but if it did, it would illustrate this point even more graphically.

Japan's current Internet topology is very similar. Each ISP radiates out from Tokyo like arms of a starfish that meet in the center. In the case of Japan's ISP system, they meet at IX's in Tokyo. This structure distorts the network in many ways.

For example, throughput between two locations in the same region will vary drastically depending on whether or not they share the same ISP. If they do not share their ISP, chances are great that any traffic between them has to be routed through Tokyo. Also, if a disaster strikes a Tokyo IX, then regional IP's will be cut off as well. As should be learned from experiences like the Great Hanshin Earthquake in 1995 or the destruction of the World Trade Center in New York, low levels of redundancy represent a clear and present danger.

Also crucial is the fact that infrastructure, capital, and human resources gravitate toward IX's. If you are going to build an ASP or data center, you naturally want to do so where a certain level of throughput is guaranteed without having to go through an ISP. From the standpoint of network topology this means a Tokyo IX. And, since this structure is self-reinforcing, economic activity beyond its actual user distribution ends up concentrated in Tokyo.

Since this dependency on Tokyo has a strong relationship with traffic levels, it is not a problem any one region can overcome on its own. Well, what if the regions teamed up? To make such an effort successful, they would need a sufficiently advanced technical grounding. This is one more rationale behind RIBB.

First we will give a summary of RIBB's activities, then explain in detail one of its most successful: video broadcasting of regional events.

2. Activity Summary

RIBB is a practical research project whose aim is to realize high-speed, broadband services and applications for end users on the next generation Internet [1–3].

The basic network connection topology of RIBB is illustrated in Figure 1. The University of Tokyo, Tohoku University, Nagoya University, and the TAO's Kochi Communication Traffic Research Center located at the Kochi University of Technology were chosen as core router sites with 600Mbps or greater bandwidth on JGN. The connection topology of the core routers is a complete graph, enabling greater redundancy and load distribution. Meanwhile, edge routers are used in other regions and organizations, which are interconnected to the project as leaf sites.

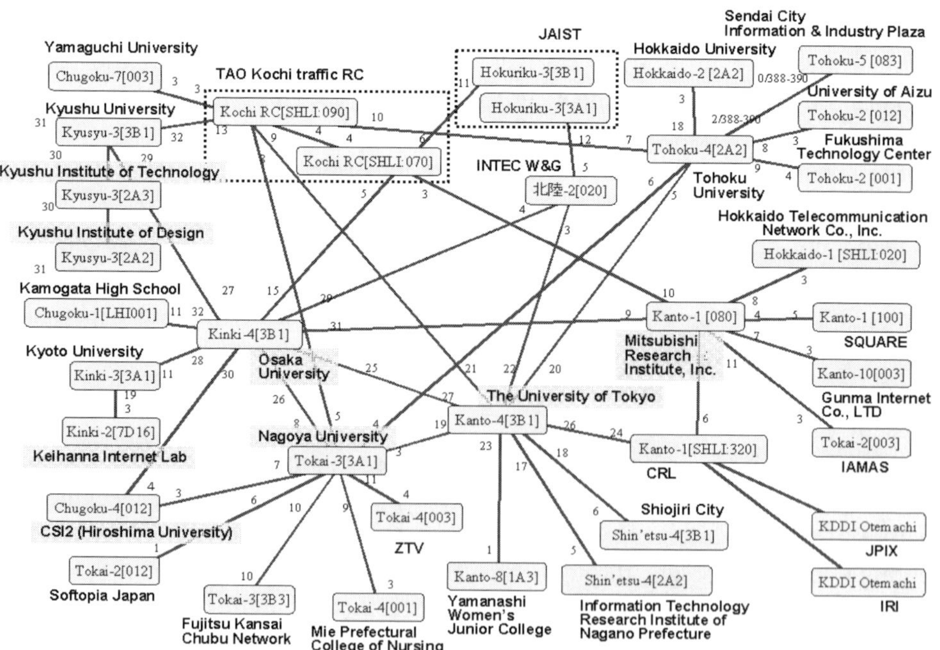

Fig. 1. Network Connection Topology of RIBB.

RIBB activities are centered around basic technology for interconnections between regions and general applied technology to make use of such connections. Three of the most important efforts are as follows.

1. Research to spur inter-regional content distribution and the synthesis of communication and broadcasting by interconnecting CATV, wireless, xDSL, and other high-speed networks through a high-speed backbone.
2. Research enabling sharing of regional information resources over a high-speed backbone, broadband distributed data centers, and other next generation applications.
3. Research on very high-speed backbone technology such as label switches and bandwidth control to implement an environment where 1) and 2) may be realized.

2.1 Research on Regional High-Speed Access Lines

Research is being conducted on high performance applications that require broadband. One example is voice and video transmissions between geographically diverse regions that are interconnected via the high-speed networks which are even now being built. In particular, the CATV (cable TV) network provides direct access to the Internet, but is also likely to exert a strong influence on the Internet through facets such as broadcasting and content technology. Thus, our project includes research on the merging of communications and broadcasting to enable streaming of CATV broadcast content onto the Internet through JGN, or even taking video content from the Internet and transforming it into a CATV broadcast.

2.2 Sharing Regional Information Resources

Individual regional IT and vitalization policies have resulted in various digital image and other content requiring high-speed Internet that are offered for use in the regions in which they originate. However, sharing of this information between regions would allow production of higher quality content through shared production costs. Also, providing information about various regional events to a wider audience, would serve a very important function from the perspective of vitalization, stimulating localities by increasing tourism.

Thus, one of our research goals is a next generation content distribution system and distributed information environment over the local high-speed networks now being constructed, which will allow sharing of local multimedia information and content, regional information resources such as information processing environments, and local content provided by CATV or newspapers.

This technology will enable expensive IT resources such as encoders and software to be shared among geographically diverse regions, allowing such locally developed resources to be leveraged for even greater effect. We are especially concentrating on content originating from CATVbecause it promises to be a crucial aspect of the merging of communications and broadcasting technologies.

2.3 Very High-Speed Backbone Technology

Backbone technology with flexible policy control and high-speed communications is necessary to perform experiments on the high performance applications described above or high-speed connections. In order to realize this, backbone architecture utilizing MPLS (Multi Protocol Label Switching) is being designed and deployed.

This activity was spurred on by outside projects becoming involved in RIBB, and developed into a next generation IX project (hereafter called Distix). For further details, please refer to the article entitled "Demonstrations of Broadband, Distributed IX Using MPLS" in this edition.

3. Video Streaming over RIBB

RIBB conducts video streaming experiments in various regions by coordinating them with events in those regions. These experiments are very close to RIBB's ultimate goals, so they possess a high level of priority within RIBB. The rationale for basing our experiments on regional events is that this kind of content helps spark interest from people far away, and because covering local events is more likely to garner cooperation from local ISP's, CATV companies, local governments, and other organizations.

Below, we will first describe some of our previous and current activities in chronological order, then go into further depth about the video transmission technology used.

3.1 Video Streaming Experiments

First, we would like to give brief summaries of all RIBB's video streaming activities in chronological order. Table 2 shows streaming content, dates, broadcast origin, and number of receiving regions. Table 3 shows the location of receiving regions. In Table 3, ☆ indicates transmission location,

Table 2. Video streaming content

No.	Content	Begun	Completed	Video source	Streaming destinations
1	Taichi Sakaiya's Lecture	2000.03.25	2000.03.25	Gifu	9
2	Toyama Athletic Meet	2000.10.14	2000.10.19	Toyama	7
3	Gigabit Network Symposium	2000.11.08	2000.11.08	Nagoya	3
4	Live! Eclipse (Total Lunar eclipse)	2001.01.10	2001.01.10	Tokyo	1
5	Kaiji Kirameki Athletic Meet	2001.01.27	2001.01.31	Yamanashi	4
6	Gigabit Network Forum 2001	2001.05.28	2001.05.28	Toyama	10
7	Live! Eclipse (Total Solar eclipse)	2001.06.21	2001.06.21	Hiroshima	9
8	IPSJ DSM SIG	2001.07.27	2001.07.27	Kochi	1
9	Sendai Tanabata Festival	2001.08.08	2001.08.08	Miyazaki	2
10	Michinoku YOSAKOI Festival	2001.09.22	2001.09.22	Miyazaki	3
11	Shin-minato Hikiyam Festival	2001.10.01	2001.10.01	Toyama	6
12	Miyagi Athletic Meet	2001.10.13	2001.10.17	Miyazaki	4
13	First Challenged Athletic Meet	2001.10.27	2001.10.30	Miyazaki	3
14	ISOM	2001.11.11	2001.11.12	Toyama	4
15	Gigabit Network Symposium	2001.11.19	2001.11.20	Tokyo	3
16	Christmas Event	2001.12.24	2002.12.24	Sapporo	1
17	Shingen-ko Festival	2002.04.05	2002.04.06	Yamanashi	3
18	Miyagi IT Forum	2002.05.21	2002.05.21	Miyazaki	5
19	Sendai Tanabata Festival	2002.08.06	2002.08.06	Miyazaki	2
20	Yosakoi Kochi Athletic Meet	2002.10.27	2002.10.30	Kochi	6
21	Yosakoi-pic Kochi, Challenged Athletic Meet	2002.11.09	2002.11.11	Kochi	6
22	Live! Eclipse (Total Solar eclipse)	2002.12.04	2002.12.04	Tokyo	6

□ indicates receiving location, and ○ indicates locations where signals were received and then re-broadcast.

1. Lecture by Taichi Sakaiya. This was RIBB's first video streaming experiment. A Sony link unit which had the ability to transmit digital video over ATM's was used. The stream was encoded as Real 20, 40, and 80kbps signal for end users.

2. National Athletic Festival in Toyama. This was RIBB's first experiment using multiple video streaming technologies and a robust streaming system.

3. Gigabit Symposium. This was the first time RIBB utilized a DVTS (Digital Video Transport System) over IP, the media layer of which used IPv4 unicast. Since in this experiment a special JGN video transmission

Table 3. Video streaming sites

No.	Sapporo	Iwate	Miyazaki	Fukushima	Tokyo	Yamanashi	Nagoya	Gifu	Mie	Toyama	Ishikawa	Fukui	Kyoto	Osaka	Hiroshima	Kochi	Fukuoka	Saga	Yamaguchi
1		○	○		○		○	☆	○	○	○			○	○				
2			○		○		○	○		☆						○			
3						○	☆			○						□			
4					☆○														
5						☆	○	○		○						□			
6							○	○	○	☆	○	○		○		○	○		
7			○		○	○		○		○			○		☆	○	○	○	
8										○						☆			
9			○					○		○									
10			☆				○			○						○			
11			○			○		○		☆							○		
12			☆			○				○						○			
13			☆							○						○			
14			○			○				☆						○			
15			○		☆					○						○			
16	☆				○														
17						☆		○		○						○			
18			☆			○		○		○						○			○
19			☆					○		○									
20			○			○	○	○		○						☆			○
21			○			○	○	○		○						☆			○
22			○		☆	○				○				○		□			○

from Kita-Kyushu to Nagoya was re-transmitted over RIBB from Nagoya, the transmission source for RIBB was Nagoya.

4. Live! Eclipse (total lunar eclipse). The Live! Eclipse project (currently Live! Universe) received video images from a total lunar eclipse in real time and re-streamed them nationwide over RIBB [3]. Since a wide audience for this broadcast was predicted, multiple content servers were prepared, and load distribution methods used to stream content to each end user's computer from the closest one. RADIX [4] and TENBIN [5] were used for this purpose, and AS paths were used to judge the distance between end users and servers. Unfortunately, even though observation locations were set up at eight locations throughout Japan, inclement weather prevented images of the lunar eclipse from being captured for broadcast at all sites except Saga, which captured only poor quality images. However, the Live! Universe project has now broadened this to include broadcasting images of solar eclipses, lunar eclipses, and meteors.

5. Kaiji kirameki National Athletic Festival in Yamanashi. This was the first experiment using an IPv4 multicast as the media layer in a DVTS. Also, because of the growing prevalence of CATV high-speed Internet access,

this broadcast was conducted at the relatively high speed of 1Mbps with Windows Media. In this experiment, transmission was first made from Yamanashi to Kochi, where it was converted to DV over ATM, and then the signal sent to Nagoya and Gifu.

6. Gigabit Forum 2001. This was the first experiment involving broadband Real (500kbps and 1Mbps) and Windows Media (1.5Mbps) multicasting over IPv4. The Real format was hardly used after this experiment. Further, at about this time restrictions on design of digital video transport topology began to grow. This is because most organizations participating in RIBB use OC3 (155Mbps) ATM's to access JGN, and, since digital video takes up about 40Mbps of bandwidth, just three digital video transmissions would monopolize the entire OC3 bandwidth.

7. Live! Eclipse (total lunar eclipse). For this experiment, RADIX and TENBIN were used to perform load distribution for transmission in Real, QuickTime, and Windows Media. Also, an MPEG2 transport system operating on IP called MPEG2TS was used to stream video to different regions, making this RIBB's first MPEG2 video broadcast.

Descriptions of the following items are abbreviated due to space considerations. The heading numbers correspond to the numbers in Tables 2–4.

1. Shinminato Hikiyama Festival. This was the first experiment broadcasting digital video over IPv6 unicast, and also the first time digital video was broadcast over IPv6 multicast.

2. Gigabit Network Symposium. Digital video from this symposium on JGN IPv6 network was broadcast over IPv6 unicast.

3. Christmas Event Live. This was the first transmission over D1.

3.2 Video Broadcasting System

The following system was used to perform broadcasting in almost all cases:

- Filming video. Sometimes a RIBB team filmed the video, and other times RIBB received video filmed by other organizations. Live video of the solar and lunar eclipses in the Live! Eclipse project are an example of the latter.
- Transmission on JGN. Broadband image transmission technology was used to transmit digital video or MPEG2 images to various regions participating in RIBB.
- Broadcasting on the Internet. Signal was converted to user-friendly formats such as Real or Windows Media, then streamed over the Internet.
- Local reception. Individuals could watch on their home PC's or at special event locations on large screens.

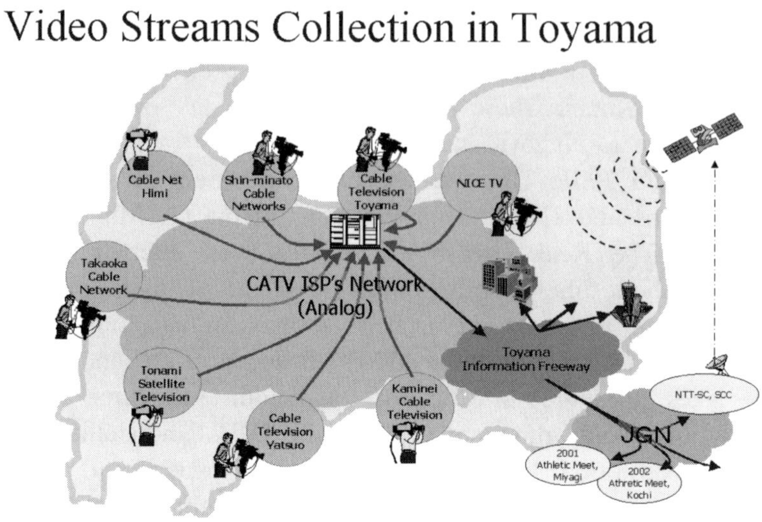

Video Streams Collection in Toyama

Fig. 2. Video Streams collection in Toyama.

This is a representative outline of the system, but many variations exist. For example, in many cases conversion to Real, Windows Media, or other formats was performed not at the transmission endpoint on JGN, but at the location of transmission origin or en route.

Broadcast System for Toyama Athletic Festival. RIBB conducted an experiment where video from the National Athletic Festival held in 2000 in Toyama prefecture was streamed live over the Internet. This was a groundbreaking event for RIBB's video broadcasting undertaking, and is described here as a representative example of our video broadcast system.

First, Figure 2 illustrates the way video was obtained for broadcast. When CATV or local television stations shoot local events, we can access local content of high quality. In this case, video shot by CATV had already been collected at one location over the CATV organization's network. Copyright and other complex issues over rights became an issue, so caution was required.

Figure 3 shows the system for streaming this content from Toyama to the rest of the country. Usually RIBB uses UBR over JGN's ATM network, but with advanced notice, a CBR can be used for event purposes.

The signal had to pass through the regions of Kochi and Miyagi to reach its final destination in some cases. This was because the limits of Toyama's JGN precluded direct transmission to all localities, so the signal had to be sent to places such as Kochi and Miyagi with extra bandwidth, from where it was re-broadcast. Figure 4 shows the re-broadcast system as used in Kochi. For this event, a link unit was used to stream the digital video directly over the ATM's PVC. Broadcasting to multiple destinations was

Video Streaming from Toyama

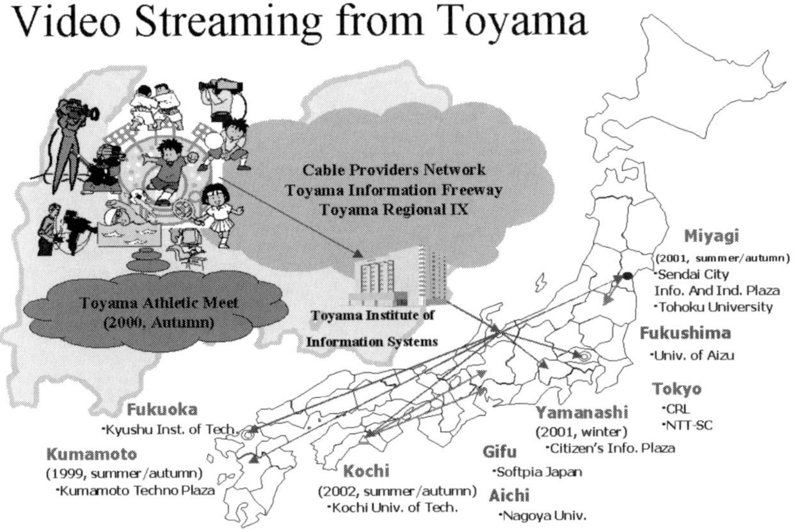

Fig. 3. Video streaming from Toyama.

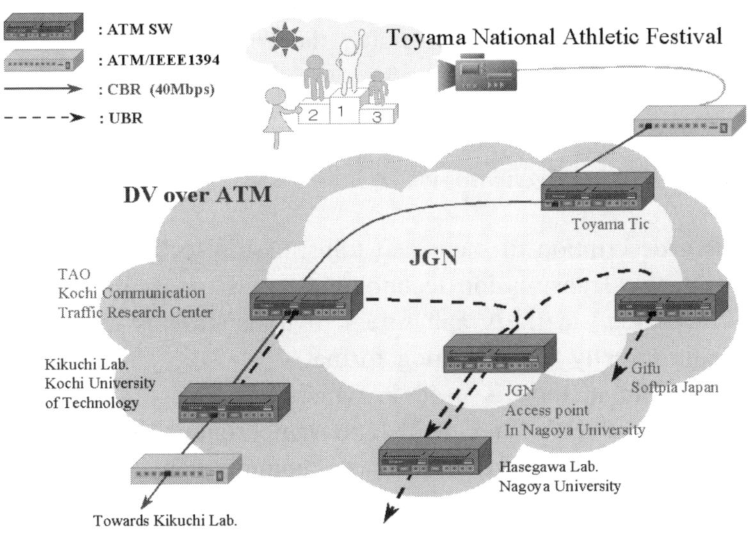

Fig. 4. Re-broadcasting with cell copying.

greatly facilitated by making cell copies at the ATM switch so that output could be broadcast to multiple PVC's. In this figure, one of the link units is referred to as "Tanbako", a common name for the unit taken from the name of its developer.

Each of the participating organizations in RIBB received the signal through JGN and then displayed the video on a special public viewing screen or streamed it over the Internet through local ISP's in Real and Windows Media formats. Figure 5 shows how this re-encoding was accomplished in Kochi.

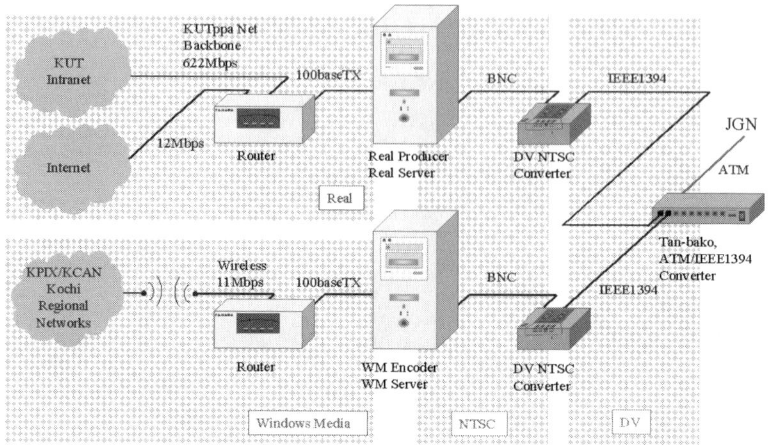

Fig. 5. Re-encoding and streaming.

After the digital video format was converted to an NTSC analog signal, it was sent to Real and Windows Media encoders connected to a video capture board for conversion. The images were then streamed in these formats to other locations within the University, the Internet, and Kochi's regional intranet KPIX/KCAN.

3.3 Video Transmission Technology

Following is a description of the video transmission technology employed for these experiments. Transmission technologies used for each event are shown in Table 4. "Analog" in the graph refers to transmissions made over CATV or other media directly in the analog format.

- D1. A recording method is used in which four image signals (brightness Y, color contrast B-Y and R-Y, and synchronous S) and four audio channels are digitized together without compression. Video quantization is performed without compression at 8 bits, with sampling frequency of 13.5MHz for Y and 6.75MHz for all other bands. Audio is quantized at 16 bits with a sampling frequency of 48KHz. These methods result in required transmission speeds of 270Mbps for video and approximately 3Mbps for audio.
- Digital Video (DV). The format used is the same as home-use DV. Inter-frame compression is not used, requiring about 30Mbps bandwidth transmission. Packet loss results in square block noise appearing over the frame being shown.
- DV over ATM (DVoA). This method transmits DV frames directly over ATM AAL5. RIBB uses a device called a link unit manufactured by Sony. Setting the ATM VC allows DV to be transmitted easily.

Table 4. Video streaming locations

No.	Video Streaming Technologies										
	D1	DVoA	DV4u	DV4m	DV6u	DV6m	MPEG2	Real	WMT	QT	Analog
1		○						○			
2		○						○	○		
3		○		○							
4			○					○	○	○	
5		○		○					○		
6		○		○				○	○		
7						○		○		○	
8			○								
9									○		
10		○							○		
11		○			○				○		
12		○							○		○
13		○							○		○
14					○	○					
15		○	○		○				○		○
16	○										
17									○		
18		○	○						○		
19		○	○						○		
20		○	○			○			○		
21		○	○			○			○		
22						○			○		

- DV over IPv4 unicast (DV4u). A DVTS developed by Keio University is used to transmit DV over UDP. IPv4 unicast is under UDP. In order to use the DVTS, FreeBSD 4.x, NetBSD 1.5.x, Linux 2.4.x, Mac OS X, or Windows 2000/XP are required. (http://www.sfc.wide.ad.jp/DVTS/)
- DV over IPv4 multicast (DV4m). DVTS is used, with IPv4 multicast used under UDP. DVMRP is used as routing protocol in most cases.
- DV over IPv6 unicast (DV6u). DVTS is used, with IPv6 unicast used under UDP. To transmit or receive signals, an installation of the IPv6 protocol suite KAME running on FreeBSD or other BSD 4.4-Lite OS, or a new OS to which KAME has been ported, is required.
- DV over IPv6 multicast (DV6m). DVTS is used, with IPv6 multicast used under UDP. A sparse mode PIM is used as routing protocol.
- MPEG2TS (MPEG2). MPEG2TS uses FEC (forward error correction), which is a method for receiver-side recovery of en-route packet loss using redundant packets[6]. This solves the problem of MPEG2's weakness in terms of packet loss and makes the standard a practical technology for video transmission. This method allows for DV equivalent quality to be obtained at transmission speeds of 6Mbps.

- RealSystem (Real). RealSystem is a live broadcasting system developed by RealNetworks comprised of the following three elements:
  - RealProducer
  - RealServer
  - RealPlayer

  This system can broadcast in either unicast or multicast over either TCP or UDP. When the number of clients exceeds 25, a license fee of about 10,000 yen per client becomes necessary. Server OS's such as Windows NT Server or Windows 2000 server are required for operation.
- Windows Media. Windows Media is a live broadcasting system developed by Microsoft. This is the most widely used software on RIBB. It's fundamental components are similar to Real's.
- QuickTime. QuickTime is a live broadcasting system developed by Apple. It requires a server OS such as Mac OSX Server.

4. Summary

When RIBB was born in 1999, it consisted of ten organizations. Now, the number has grown to thirty-five, which may be the largest number of organizations participating in any JGN project. This illustrates the powerful latent needs and desires for regional Internet activity, and how these have found an outlet through a remarkable alliance with JGN.

One could say that JGN does nothing but transport ATM cells. A different perspective shows us that it has the capability to unify what had been disparate regional elements, making it possible for instance to broadcast video to potentially the whole of Japan. Most important of all, however, the knowledge and experience which had lain isolated in Japan's various regions have now come together under the roof of RIBB.

In the age of UUCP, in which electronic mail and news was exchanged over dial-up connections, it was natural for organizations physically close to each other to link up and form their own regional networks. When this technology gave way to IP connections over dedicated lines, traffic increased and volunteer-based operations were stretched to their limits. Afterwards, commercial ISP's appeared, and one could pay for whatever services were needed. Finally, the Tokyo-centric network topology we see today developed.

Now, however, a groundswell of regional activity has begun, and the Tokyo-centric structure has begun to give way. Before RIBB got underway, we were at the Internet Research Institute, where Dai Nishino said to us "We're going to change the whole design of Japan's Internet architecture". When we look at things now, it appears as though that ambition is truly coming to pass.

In the future, we will share the know-how we have gained up to the present by making our experiences, analysis, considerations, and results public. And, by developing applications that make use of regional content, sophisticated techniques for sharing it, and multi-regional redundancies, we would like to develop the idea of regional interconnections into a virtually ubiquitous one.

Acknowledgements

RIBB is supported by its many participants, not all of whom it is possible to name here. Therefore, we will confine our individual thanks here to those who directly aided us in the preparation of this manuscript. The video broadcast illustrations were received from Eikoh Ito'oka at the Toyama Institute of Information Systems and Michiko Sugiyama, who at the time was on the RIBB team of the Professor Kikuchi's laboratory at Kochi University of Technology. The summary of video broadcasting activities is based on a survey by Ken'ichi Kanayama of the Intec Web and Genome Informatics Corporation. We would also like to express our gratitude to everyone involved in regional video content creation, including "Live! Universe" project and all the different regional CATV organizations.

RIBB is officially called TAO JGN-G11012. The people at the TAO, which is a special corporation under the Ministry of Public Management, Home Affairs, Posts and Telecommunications, have also done everything possible to make our project a success, for which we offer our heartfelt thanks.

Finally, we would like to express our gratitude to everyone else whose invaluable contributions have made the project a success.

References

[1] Nakagawa, I., Hayashi, E., Hiji, M., Yatsushiro, K., Kikuchi, Y. and Nishino, D.: A Trial for Interconnection on the Gigabit Network, Information Processing Society of Japan Research Report 99-DSM-15, pp. 7–12 (1999), ISSN0919-6072.

[2] Kikuchi, Y., Nakagawa, I., Hiji, M., Yatsuhiro, K., Nishino, D., and Hayashi, E.: A Trial for Reconstructing the Ground Design of the Internet Architecture in Japan, Proceedings of the Second International Conference on Advances in Infrastructure for Electronic Business, Science and Education on the Internet (SSGRR) (2001).

[3] http://www.live-universe.org/

[4] http://www. toyama.net/~ikuo/

[5] http://www. tenbin.org/

[6] http://net.ipc.hiroshima-u.ac.jp/mpeg2ts/

Chapter 5

The JB Project

Hiroshi Esaki[a], Akira Kato[a],
Hideo Miyahara[b] and Jun Murai[c]
[a] The University of Tokyo;
[b] Osaka University; [c] Keio University

Abstract. The JB project is a research and development project which was initiated in 1998 centered on the WIDE project (http://www.wide.ad.jp), the ITRC (Internet Technology Research Committee, http://www.itrc.net/), and the CKP (Cyber Kansai Project, http://www.ckp.or.jp/). The project's goal is the establishment of a high performance, broadband Internet architecture and technology with a research and development testbed through practical operation and experimentation (headed by Professor Jun Murai of Keio University). The JB project also collaborates with other international organizations such as Internet2, APAN, AI3, and APII, and functions as an internationally advanced R&D testbed for international cooperative research over high speed international connections.

1. Introduction

Internet technology uses a wide variety of data links to build a global and ubiquitous digital communication infrastructure. It has evolved from a digital network intended solely for the computer scientists of the time to one which interconnects ordinary people (who may not even familiar with information technology) and digital devices from all over world and contributes to both industry and private activities. As the functionality required from the Internet becomes ever more diverse in the 21st century, it will have to become ever more flexible to satisfy the needs of a growing array of digital devices and usage demands. In other words, the Internet has to evolve successfully into a global ubiquitous digital network infrastructure for the 21st century. In order to achieve this evolution, the JB project offers to the researchers and engineers connected to it an environment where they can perform necessary R&D activities, real-world testing over an actual and practical network, and advanced networking. This R&D environment helps researchers and engineers to clarify and understand the key technologies for making the next generation Internet a reality and speeds the R&D of new applications.

The next generation Internet has to provide the following kind of environment:

- Internet for everyone
- Internet for everything
- Internet everywhere
- Internet at any time
- Internet anyway

After 1998, the JB project integrated all of its strategic R&D efforts into IPv6, the protocol for the next generation Internet infrastructure. With 128bit addressing, IPv6 will not only support the massive address space that will be required of the next generation Internet, but will also preserve the end to end transparent digital communications infrastructure of the original Internet. In other words, IPv6 will maintain the "end to end architecture" which has guided the development and evolution of the Internet thus far to be brought forward into the 21st century.

Some of the middleware, applications, and IPv6 infrastructure software itself which have been developed and tested in real-world use on the JB project's IPv6 network are also used in academia and industry, including internationally, where they have secured a reputation for quality and reliability through wide use and have become the global de-facto standard.

In this chapter, I would like to give an overview of JB project, R&D on infrastructure technology and applications, and the iGrid 2000 and IETF 2002 (Internet Engineering Task Force) events, which the JB project both participated in and provided technical support for. Then I will give a description of the current network structure of the JB project's test-bed in terms of domestic and international connections, and finally future research items as well as directions for development of the experimental network.

2. Summary of the JB Project

The purpose of the JB project is to build and operate a nation-wide testbed for next generation R&D as a technology testing and evaluation environment. One more crucial aspect of the JB network is operating the JB network as an important autonomous network in a global R&D testbed along with other international research organizations such as APAN, APII, AI3, Internet2/Abilene, and 6REN [1–3].

The WIDE project, ITRC, and CKP have all cooperated in the JB project testbed architecture, which utilizes various data link technologies as well as interconnections and overlays with multiple organizations.

JGN (Japan Gigabit Network)

Communications Research Laboratory, APII testbed (http://www.apii.network/)

APAN/TransPac (http://www.apan.net/)

WIDE Internet (including joint research by NTT Group and PowerdCom/ Ttnet)

AI3 testbed (http://www.ai3.net/)

BBCC (as of 9/2002)

2.1 R&D on Infrastructure Technology

2.1.1 IPv6 Technology

The JB project has maintained and expanded the R&D activities on IPv6 which the WIDE project has been pursuing since the mid 1990's. The WIDE project has performed three major special projects for IPv6 referenced protocol stack development: (1) The KAME project (http://www.kame.nct), which performs R&D on IPv6 protocol stacks for BSD types of Unix, (2) the USAGI project (http://www.linux-ipv6.org/), which performs R&D on IPv6 protocol stacks for Linux, and (3) the TAHI project (http://www.tahi.org), which performs R&D on IPv6 node testing specifications and software (Figure 1 shows logos for each project).

The IPv6 protocol stacks which are the results of the KAME project's R&D have been integrated with those developed and deployed independently by the INRIA of France and the NRL (Naval Research Labs) of the USA, and are now in global circulation as the built-inprotocol stack in BSD Unix.

Meanwhile, the results of the USAGI project's R&D have begun to be adopted in the main Linux source code tree as a patch program since April, 2002. Requests for cooperation have also been made to the TAHI project from the PLUG group of Europe's ETSI and the Connectathon group in the USA. TAHI has also already hosted three interoperability test events in Japan, with many participants from overseas as well as Japan.

The KAME project has lead to the realization that the porting of routing protocol suited to IPv6 is an important technical issue for the deployment of IPv6 technology. Development and integration with KAME's kernel as well as RIPng and BGP4+ have been completed in the early stage of the KAME project. Now, we have begun focusing on R&D on OSPFv3, an interior

Fig. 1. IPv6 stack R&D projects.

gateway protocol (IGP) used on large scale autonomous networks. The JB project's R&D on OSPFv3 is conducted in collaboration with IP Infusion of the USA (http://www.ipinfusion.com/) with daily operation on both KAME and USAGI platforms on the JB network.

Next, in order to truly implement an IPv6 system on a global scale, the DNS (Domain Name Server) system, which offers a directory service on a global scale for FQDN (Fully Qualified Domain Name) and IP addresses, must be enabled for IPv6. The WIDE project bears responsibility for operating one of only thirteen root DNS servers (M root servers) in the world. The DNS system is a client-server type system, in which clients normally use a software module called "resolver" and servers use one called a "bind". "Bind" stands for Berkeley Internet Name Domain, and is distributed freely through the ISC (Internet Software Consortium, http://www.isc.org/). The JB project (WIDE project) continues a joint effort with the ISC to port "bind" to IPv6. It is also involved in research activities to port the DNS system to IPv6. The project especially bears responsibility for porting the root DNS servers to IPv6, including the server location of root DNS servers, packet transmission methods, and migration scenarios.

2.1.2 Network Monitoring Technologies

The JB project continues research and development of a network monitoring system based on SNMP (Standard Network Management Protocol) which will (1) efficiently collect various network management information needed by a wide variety of users and applications and (2) present that information effectively.

Continuous network management information collection from multiple nodes using SNMP involves challenges such as synchronization of information collection times, reduction of fluctuations in collection time intervals, and reducing traffic incurred by information collection activity itself. Since December, 1999, the JB project has performed remote data collection and accumulation using SNMP while at the same time conducting R&D efforts on a system which can meet these challenges (Figure 2). For these experiments, network management information was collected at each of the routers making up the JB of in/out IPv4 traffic levels for each router, in/out IPv6/4 traffic levels for each IF, and status and other information.

Next, research on expanding the SNMP protocol to realize seamless management was carried out for the purpose of monitoring IPv6/4 dual stack networks. Specifically, this included research on SNMP transport protocols on IPv6/4 dual stack networks, porting SNMPv1/v2/v3 to IPv6, and research on multi-protocol mutual conversion agents, each of which has been actually deployed [4].

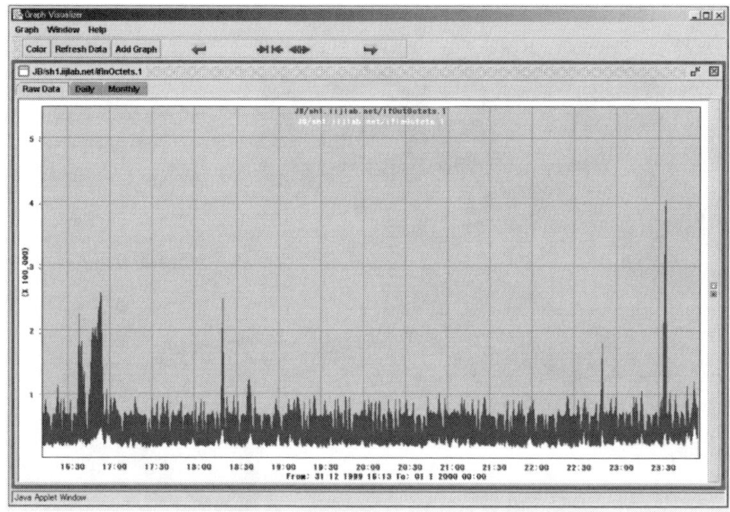

Fig. 2. Traffic measurement.

2.1.3 Multicast Technology

The JB project is pursing research on large scale multicast technology using PIM (Protocol Independent Multicast) [5] and XCAST technology for multicasting to small groups of people, as well as multicasting technology which can reduce error rates using FEC (Forward Error Correction).

1. Large Scale Multicasts Using PIM. On Saturday, November 27, 1999, digital video stream taken from the WIDE project's research meeting held at Kurashiki University of Science and the Arts (in Okayama prefecture) was broadcast using PIM-SM (PIM-Sparse Mode) over the JB network to about 10 of the primary sites of the JB project in real time as a one-way broadcasting experiment (Figure 3). In March of 2002, a Japan-US teleconferencing demonstration was held with cooperation from Internet2/Abilene/, the Fujitsu Laboratories of America in Maryland, and NTT Communications MCL in Palo Alto, USA. The project is also performing R&D on SSM (Source Specific Multicast) and studying how to introduce this to the experimental network.

2. XCAST (Explicit Multicast, http://www.www.alcatel.communication/xcast/). XCAST is a multicast format which, instead of using the group addressing of the traditional ISM (Internet Standard Multicast), allows for recipient designation by explicitly writing multiple multicast addresses in packet headers. XCAST has already been officially proposed as an Internet-Draft to the IETF jointly by IBM, Alcatel, and others as MD06 (Multicast Destination Option on IPv6). This method boasts exceptional scalability for group numbers, and will be especially suited to and effective for the

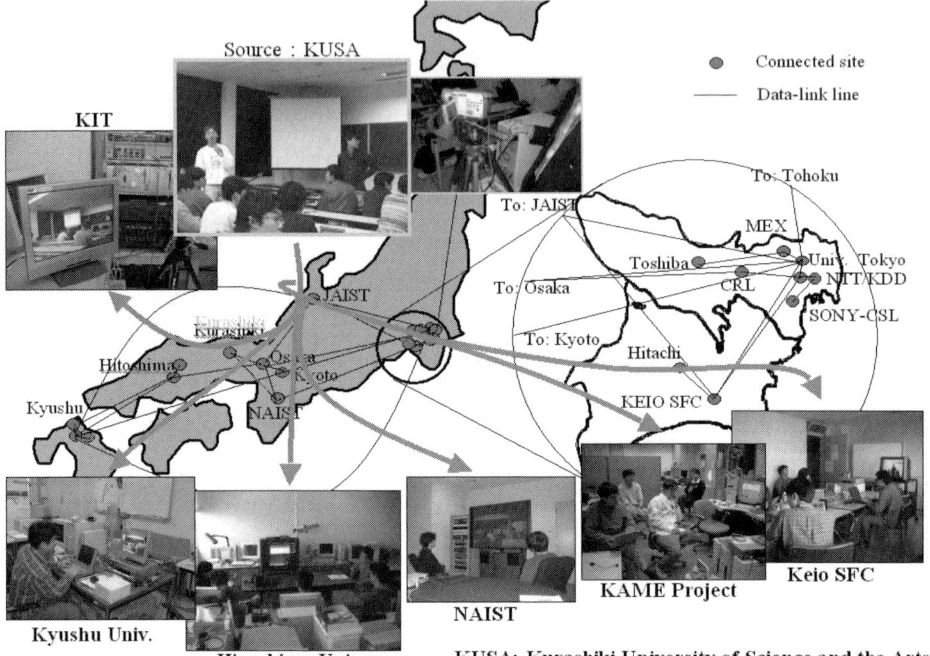

Fig. 3. IPv6 DV multicast experiment using PIM-SM.

numerous small private groups wishing to create their own multicasts, from teleconferencing for organizations to networked games. Using IPv6's Destination Option, the IP addresses of all of a multicast group's nodes may be explicitly written in all IP packets. Unfortunately, at the present it is not on the IETF's standardization track, but experimental operation of the system is underway with organizations from Korea and other countries, where technology testing and development of new functions is continuing.

2.1.4 DVTS (http://www.sfc.wide.ad.jp/DVTS/)

Digital Video (DV) is a consumer-use high resolution digital video standard which uses intra-frame compression for images and PCM (12 or 16bit) for audio to deliver television quality (720×480 pixels) images. DV also uses an IEEE 1394 interface to transfer data between video devices and computers. Video coding methods in DV use intra-frame predictive coding methods only, making each frame independent because of the lack of inter-frame predictive coding.

Except for DV, all international video coding standards use Motion Compensation (MC) of MC – DCT (Discrete Cosine Transform) technology, a method using motion prediction between frames to increase compression efficiency. DCT (Discrete Cosine Transform), on the other hand, is an intra-frame coding technology which is also used on still images. Because DV uses DCT alone, it isn't as efficient as other international coding standards such as MPEG; however, it does allow high-speed encoding and decoding without special hardware and allows for easy editing of recorded video.

We have encapsulated the DV signal in IP packets in our R&D of the Internet transmission system DVTS (Digital Video Transmission System). In this system, the transmitter of the data transmits DV video and audio data from their recording device to PC using an IEEE 1394 interface. The computer then converts the DV data to IP packets (DV over RTP over IPv4/6) for transmission over the Internet. The computer which receives the data encoded as IP packets outputs it through an IEEE 1394 interface to a Digital Video deck or other DV equipment, where it is displayed directly on the computer display.

In cases where there is insufficient bandwidth for DVTS (it usually requires about 35Mbps), the system compensates for the decreased transmission rate by taking advantage of the intra-frame coding and controlling the frame rate. The system performs no special processing on audio signals.

At the beginning, DVTS only operated on BSD Unix, and video could only be output to external displays. Recently, however, it has been ported Linux, Windows, and Macintosh platforms, and has even been made capable of displaying directly on the receiving computer itself (including display on X-Windows systems).

2.1.5 Traffic Measurement Technology

R&D on traffic analysis and measurement is performed mainly on the WIDE project through MAWI (Measurement and Analysis on the WIDE Internet) working group. One landmark was the development of AGURI (Aggregation-based Traffic Profiler) in 2001, a tool in actual use on the network which efficiently collects traffic measurements over long periods of time and characterizes this traffic in terms of patterns [6]. This software both characterizes flow trends and monitors traffic over both long and short periods by using Patricia tree algorithms and the LRU (Least Recent Used) method to maintain data levels at a specified level, thus enabling storage of large amounts of data (Figure 4). Some uses of AGURI are detection of DoS attacks and traffic engineering.

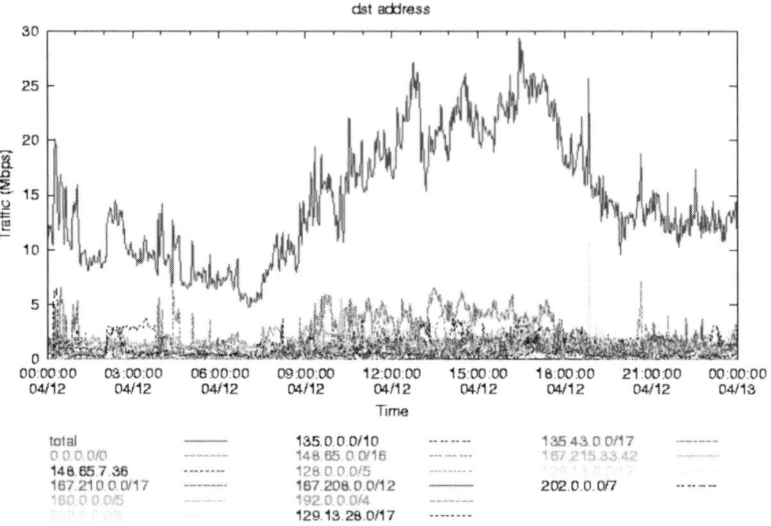

Fig. 4. Traffic measurement results using AGURI.

2.1.6 Packet Scheduling Technology

At the time the JB project was started, one of the main R&D goals was offering QoS/CoS using Diff-Serv and Int-Serv technologies. In the course of pursuing R&D on QoS/CoS, it become clear that many policies besides these existed, and that QoS/CoS had to be viewed as just one example of policy control technology. A result of this is that policy control technology is also now being researched as an extension of QoS/CoS R&D. The goal of this research team is to develop a system architecture which can establish both versatile and highly abstract policy definition that can enable policies from various users (both system administrators and end users) flexibly and without conflicts.

Currently, real-world testing is being carried out on a policy control system for offering QoS/CoS using a Bandwidth Broker system, one of the possible directions policy control technology can take.

Packet scheduling functionality is enabled by the ALTQ system [6], which was developed as part of the WIDE project. The ALTQ system integrates multiple packet scheduling mechanisms (including CBQ and WFQ) as well as multiple packet discarding mechanisms (such as RED). ALTQ has also been fully integrated into the IPv6 system software of the KAME system, making Diff-Serv possible in an IPv6 environment. The ALTQ system is used not only in Japan, but also in multiple research organizations in the USA conducting R&D on Diff-Serv and Q-Bone (End to End Quality of Service Backbone), as well as in the rest of world.

2.1.7 Mobile Technology

1. LIN6 (Location Independent Networking for IPv6, http://www.lin6.network/). LIN6 is an IPv6 compliant LINA (Location Independent Network Architecture) which separates node identifiers and node locators. This protocol allows for communication to continue even if the node moves, allowing end to end node authentication and identification. It also easily allows multihoming. The IETF is standardizing Mobile IPv6, but LIN6 is superior in many respects, including lack of header overhead and robustness.

2. GLI (Geographical Location Information). GLI is middleware for managing location information for mobile entities connected to the Internet. Mobile devices register their present location with GLI, allowing searches for the location of these entities to be performed over GLI servers. There are two types of search request: one in which the mobile entity's identifier is input and the location returned, or one in which a location range is input and the identifiers of all entities within that range are returned. Distributing the servers hierarchically allows for a large scale operation on a global basis. Privacy concerns are also reflected in the system: mobile entities may only be searched by identifier by searchers to whom they have granted such authority; other searchers may access publicly available information only.

3. Network Mobility. Research is also being performed on supporting mobility not only of nodes but of networks, with standardization presently being debated within the IETF. Fundamentally, this will entail expanding the Mobile IP architecture to network mobility.

Some elemental technologies that are the results of this research are being considered for implementation in various projects, including an "Internet Automobile".

2.1.8 Authentication Technology

Development of authentication technology now in use in Internet environments for both individual users and network devices began in the latter half of the 1990's. One current trend is the increasingly wide use of authentication technology based on X.509. As a response, the WIDE project (moCA WG) initiated the WIDE ROOT CA in September, 1998, as a root CA.

The WIDE ROOT CA is a CA for performing certificate authentication for individual CA's offered on the WIDE project such as moCA and SOI CA. One purpose of this project is the establishment of an operating authentication system technology by establishing operation technologies for root CA, insuring interoperability with various applications, and accumulating experience in root and lower layer CA operations by performing root CA key exchange experiments.

Individual experiments on moCA and SOI CA continued after the WIDE ROOT CA key exchange in January, 2000. Software used in operation of WIDE ROOT CA is also being developed, and in fiscal year 2001 the group participated in the "Challenge PKI" multiple CA interoperability testing project sponsored by the Japan Network Security Association (JNSA). By participating in this experiment project, the group has made progress in understanding the obstacles to and developing solutions for bridging, cross certification, and other technologies which will enable increased flexibility in CA operation environments.

2.1.9 UDLR

UDL stands for "Unidirectional Link", of which satellite connections are a prime example of UDL. UDLR stands for "Unidirectional Link Routing" over networks.

Most routing technology currently in use on the Internet assumes the existence of bi-directional communication links. Thus, the UDLR working group began activities in 1999 as an IETF routing area work group. AI3, CRL, and JCSAT are collaborating with the WIDE project in undertakings such as constructing and operating experimental networks in Asia using UDLR technology. The INRIA of France is also cooperating in pushing international standardization with R&D of UDLR (Figure 5). The UDLR

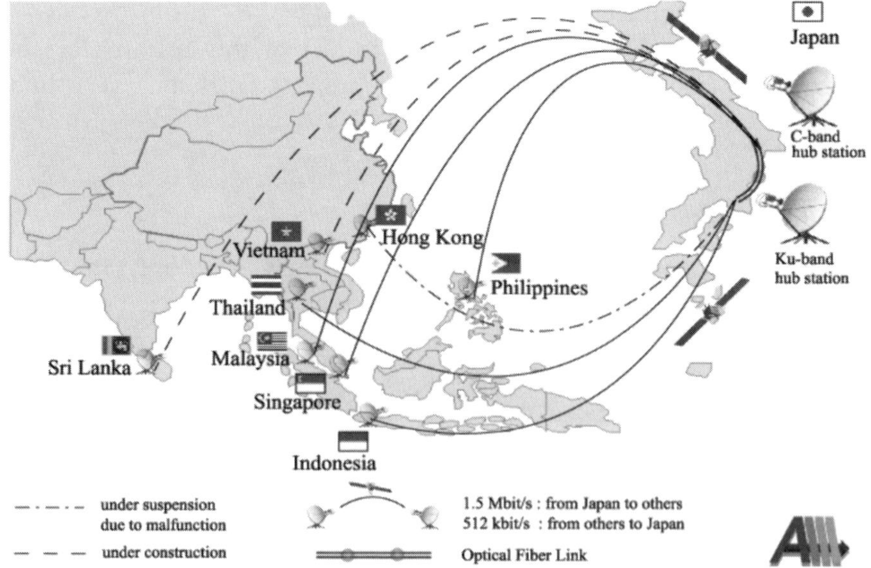

Fig. 5. AI3 network using UDLR technology.

system has progressed into the operations phase of multicasting with IPv6 and PIM.

2.2 Applications

2.2.1 Remote Electron Microscope Collaboration Project

This is a project in which the Ultra High Voltage Electron Microscope (UHVEM), one of the world's largest, at Osaka University, is controlled remotely by researchers at University of California San Diego's NCMIR (School of Medicine's National Center for Microscopy and Imaging Research) in real time and interactively. Transmission of video in real time with quality of at least DV level must be accomplished in order to assure efficient remote operation. The JB project network, ΛPAN, Abilene, vBNS, and SDSC (San Diego Supercomputer Center) all cooperated in network architecture and operations for this project, which has been designated an iGrid project.

2.2.2 SOI (School of Internet) Project

The SOI is devoted to establishing the architecture of a virtual university learning environment over the Internet. Since 1997, the WIDE project has performed R&D activities for this project, including practical test-bed operation. The SOI environment reached the next stage of R&D, called as GIOS, in 1999. The GIOS project is a large scale, high performance research experiment integrating network environments with next generation network technology. This GIOS experimental environment contains material from multiple university campuses and classes and is designed based on a near-future Internet environment including high-speed home access, dispersion of service areas, and global reach. In the first GIOS project experiment carried out in the first half of 2000, the usefulness of this design was evaluated in an integrated real-world operation across seven universities and 100 residences in Japan and the USA by utilizing next generation technologies developed by the WIDE project such as IPv6, multicasting, satellite communications, high-speed DV transmission, and classroom applications.

The GIOS project is also developing in new directions such as SOI ASIA, which is in the process of integrating with AI3 and spreading across Asia, and SOI Studio, a project run by Fujitsu Laboratories of America in Maryland and NTT Communications MCL in Palo Alto, USA.

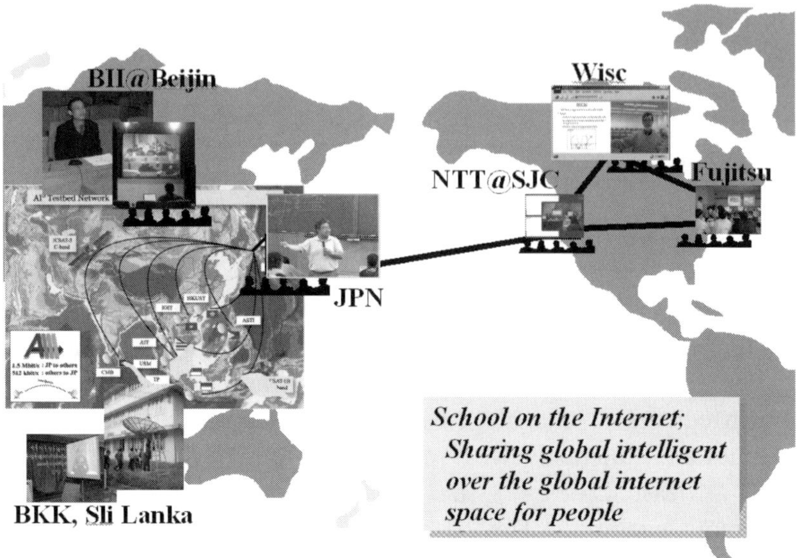

Fig. 6. Global SOI experiment network.

All of these projects include multiple directions and many goals, such as the testing and evaluation of elemental network technologies in an integrated environment, international contributions such as technology transfer and overseas teleconferences, and, finally, the strengthening of international cooperation (Figure 6).

In teleconferencing, for example, DVTS and SMIL (Synchronized Multimedia Integration Language) were integrated into a multimedia presentation system in which independent multimedia objects could be used as a synchronized multimedia presentation system.

2.2.3 IAA (I Am Alive) Project

This research focuses on issues such as the role of the Internet in society as social infrastructure, how it can be used in disasters, and what challenges lie in hardening the system against such disasters. The IAA system utilizes computers and networks to collect information and offer searchers on survivors. In order for IAA to function adequately in case of disasters, it will have to offer superior redundancy and stability. Security and privacy must also be considered for whatever services the system will offer. Further, designing scalable, robust software and network usage methods must also be given full consideration from the system performance point of view. This project, which was spurred by the January, 1995, Great Hanshin Earthquake, is continually pursing R&D into elemental technologies to satisfy these system requirements, system architecture technologies, and finally system operation technologies.

2.3 International Events

2.3.1 iGrid (http://www.startap.net/igrid)

iGrid is an international collaboration for the R&D of scientific applications using high speed, high performance Internet infrastructure. The main iGrid2000 executive committee consists of members from the EVL (Electronic Visualization Laboratory) at Indiana University, Illinois University, Argonne National Laboratory, University of Tokyo, and Keio University. iGrid 2000 has already performed a total of twenty four demonstrations of revolutionary applications over high-speed networks, including worldwide distributed computing, remote medicine, remote operations, remote collaboration, high-resolution digital multimedia collection and transmission, and real time transmission of D1 video over IPv6. In order to conduct these demonstrations, an international demonstration system architecture was developed and operated with the cooperation of the JGN, APAN TransPac (Asian Pacific Advanced Network Trans Pacific), START TAP, WIDE project network, PNJC, NTT Communications, and TTNET (Figure 7).

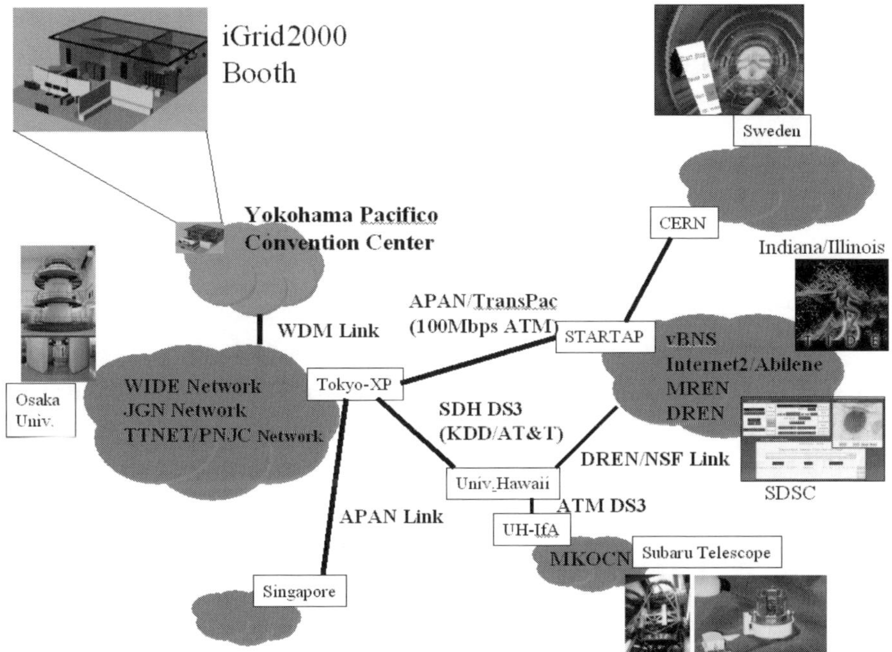

Fig. 7. iGrid system structure.

2.3.2 IETF 2002

The 54th IETF was held at Pacifico Yokohama from July 14–17, 2002. The WIDE project served as the conference's host, with help from Fuijitsu and many other sponsors which made the event a success. Just two weeks prior to the event, the FIFA World Cup Finals had been held in Yokohama, and an IPv6 based wireless LAN system had been established between Narita International Airport and the convention center (Figure 8). Almost 90% of the participants made use of this wireless LAN connection, which made the age of wireless networking feel closer to everyone.

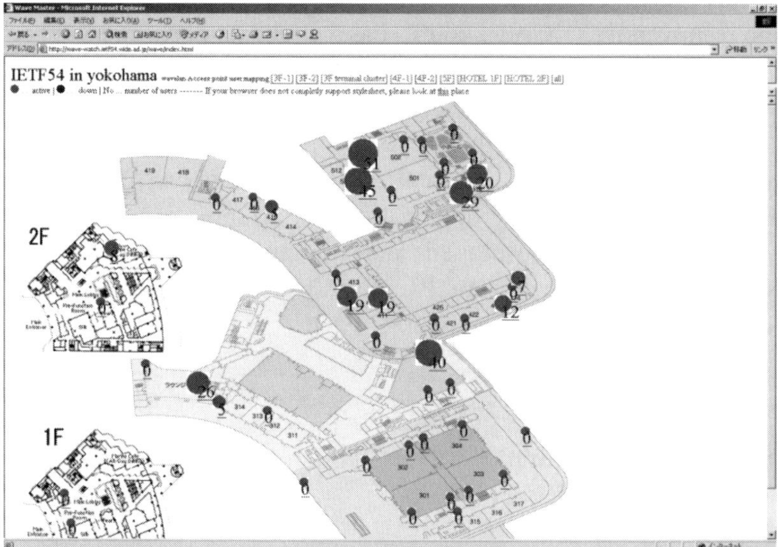

Fig. 8. Wireless LAN usage at IETF.

3. Network Structure

3.1 JB Project Internet

Figure 9 shows a simplified diagram of the structure of the JB project's Internet backbone structure. Most of the Layer 2 links are JGN ATM links. WDM links, SDH links, and satellite links are also involved, all combining to create an IPv6/4 network architecture.

3.2 External Connections

Most external connections are conducted over NSPIXP2/3/6. NSPIXP6 is an IX (Internet eXchange) dedicated to IPv6 and is where most network testing

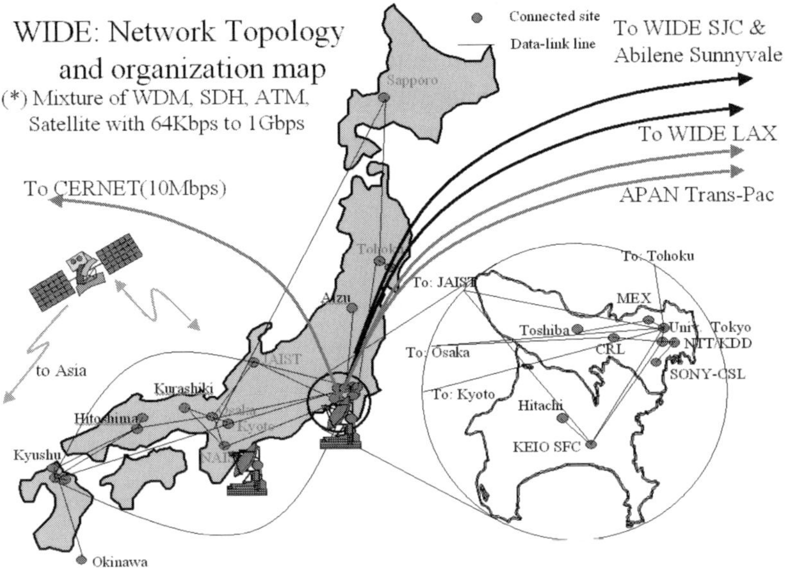

Fig. 9. Structure of JB Project's internet backbone.

and IPv6 peering with commercial providers takes place. Interconnection with the US R&D network Internet2/Abilene is performed through the APAN Tokyo-XP and Abilene Sunnyvale POP.

The network also maintains interconnections with other Asian entities, such as the Chinese R&D network CERNET, with which it uses tunneling to perform IPv6 peering (10Mpbps ATM link). IPv6 (satellite UDL) interconnections are exist with other Asian countries using the satellite antennas at Keio University and the Graduate School of the Nara Institute of Science and Technology.

4. Summary

The JB project pursues the R&D of next generation Internet technology through cooperation with the WIDE project, ITRC, and CKP while also operating an R&D testbed which delivers practical outputs via collaboration among academic and industrial groups. By developing cutting edge technologies such as multicast technology, QoS service architecture/technology, and next generation middleware and application technology, the project allows individual technologies to be integrated, evaluated and operated experimentally on the active JB network. In the SOI (School of Internet) project, especially, all cutting edge technologies have been integrated using IPv6.

In the future, the JB project will form joint collaborations with even more international research organizations in its continuing effort to advance research on next generation Internet technology. One especially important challenge is the development of an infrastructure (RF-ID, networked appliances, locators, etc.) for a ubiquitous networking environment and applications for use in such a system.

References

[1] http://www.internet2.edu/
[2] http://www.ucaid.edu/abilene/
[3] http://www.6ren.net/
[4] http://www.www.net-snmp.org/
[5] McCloghrie, K., Farinacci, D., Thaler, D. and Fenner, B.: Protocol Independent Multicast MIB for IPv4, IETF RFC2934 (October, 2000).
[6] http://www.csl.sony.co.jp/person/kjc.html
[7] Yoshida, K., Mori, H., Shimojo, S., Kadobayashi, Y., Akiyama, T. and Ellisman, M.H.: Design of a Remote operation System for Trans-Pacific Microscopy via International Advanced Networks, Journal of Electron Microscopy 51 (Supplement): S253-S257 (2002).

Gigabit Network
T. Saito and H. Esaki (Eds.)
Ohmsha/IOS Press, 2003

Chapter 6

Project for the Research of Application Technologies for Next Generation Broadband Networks (GENESIS)

Isao Kaneko[a], Fumito Kubota[b], Hideo Miyahara[c],
Shinichi Nakagawa[b], Yuji Oie[d] and Shinji Shimojo[c]
*[a] SCC Inc.; [b] Communications Research Laboratory;
[c] Osaka University; [d] Kyushu Institute of Technology*

Abstract. This project is dedicated to the research and development of fundamental technologies necessary for the architecture and use of next generation high-quality broadband networks (the next generation Internet). These networks will function as the infrastructure for the coming global multimedia society.

Efficient multimedia communication in real time will be indispensable for the next generation Internet, but varied elemental technologies will need to be combined effectively to make this a reality. Cooperation and interoperability on an international basis are also crucial to such an undertaking. This project carried out extensive research ranging over eight theoretical and practical sub-topics under the three main topics of network control, applications, and measurement technologies. We also interconnected with other international broadband R&D networks and actively collaborated in international joint experiments. This chapter outlines the project's structure and introduces the R&D carried out under its auspices.

1. Introduction

The Project for the Research of Application Technologies for Next Generation Broadband Networks was inspired by the rapid spread of the Internet and the worldwide movement toward multimedia societies. Thus, the project was started in 1997 as a TAO (Telecommunications Advancement Organization) research project under the name of GENESIS (Global Experimental Networks for Information Society) with the goal of collaborating with many countries on solutions for the technological problems inherent in the realization of

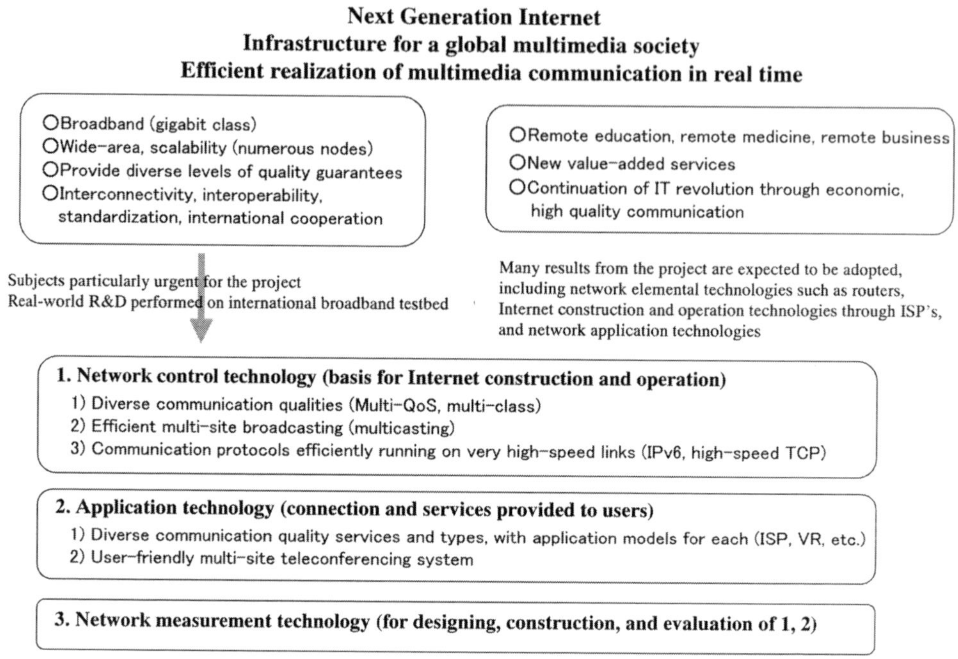

Next Generation Internet
Infrastructure for a global multimedia society
Efficient realization of multimedia communication in real time

○ Broadband (gigabit class)
○ Wide-area, scalability (numerous nodes)
○ Provide diverse levels of quality guarantees
○ Interconnectivity, interoperability,
   standardization, international cooperation

○ Remote education, remote medicine, remote business
○ New value-added services
○ Continuation of IT revolution through economic,
   high quality communication

Subjects particularly urgent for the project
Real-world R&D performed on international broadband testbed

Many results from the project are expected to be adopted,
including network elemental technologies such as routers,
Internet construction and operation technologies through ISP's,
and network application technologies

1. Network control technology (basis for Internet construction and operation)
   1) Diverse communication qualities (Multi-QoS, multi-class)
   2) Efficient multi-site broadcasting (multicasting)
   3) Communication protocols efficiently running on very high-speed links (IPv6, high-speed TCP)

2. Application technology (connection and services provided to users)
   1) Diverse communication quality services and types, with application models for each (ISP, VR, etc.)
   2) User-friendly multi-site teleconferencing system

3. Network measurement technology (for designing, construction, and evaluation of 1, 2)

Fig. 1. Relationships among R&D subjects.

such a global multimedia society. This initial phase of the project, now termed Phase I, concluded in March 2000, but its research was continued under Phase II from April 2000 to March 2002 after meeting many of its research goals. This chapter focuses on research carried out in Phase II of the GENESIS project.

The GENESIS project began with the realization that the integration of various broadband communication technologies on the Internet – incorporating not only text but also audio and video information – had already begun creating a multimedia society. However, in order to truly realize new multimedia services meeting a variety of demands, not only a broadband network architecture but also application technologies would be the key to the next generation Internet. A great number of elemental technologies are necessary for a broadband, high quality next generation network, for which just as great a number of R&D projects are underway. The project discussed herein had three main research themes which were pursued in tandem with one another. These are "control technology", which will provide fundamental architecture and operational technologies for next generation broadband networks, "application technology", which will offer multimedia communications and services to users, and "measurement technology", which will be necessary for the design, implementation, and evaluation of the first two technologies (Figure 1). Below we outline summaries each of

Table 1. GENESIS Phase II research structure

Participants	Affiliation
Project Leader Hideo Miyahara	Joint Research Institutions
Kita-Kyushu Next Generation Broadband Network Research Center	Kyushu Institute of Technology
Subleader Yuji Oie, Researcher Kazumi Kumazoe, Research Fellow Takeshi Ikenaga, Research Fellow Yoshiaki Hori	
Subleader Isao Kaneko, Researcher Masato Tsuru, Research Fellow Kenji Kawahara	
Toyonaka GENESIS Research Group	Osaka University
Subleader Shinji Shimojo, Researcher Hikaru Yagi, Research Fellow Naoki Wakamiya, Research Fellow Go Hasegawa, Research Fellow Kazutoshi Fujikawa, Research Fellow Takeshi Okuda	
Koganei GENESIS Research Group	CRL KEK
Subleader Fumito Kubota, Researcher Shuji Uetsuki, Research Fellow Yukio Karita	
Subleader Shin'ichi Nakagawa, Researcher Yoshiaki Kitaguchi	

these research areas and the broadband experiments and trials which were conducted on them both domestically and in international collaborations.

The research organization (Table 1) is divided into three groups located in Kita-Kyushu, Toyonaka, and Koganei. The JGN (Japan Gigabit Network) was used to conduct frequent teleconferencing between these groups over DV (Digital Video). Each of these research groups conducted joint research in close collaboration with the Kyushu Institute of Technology, Osaka University, Communications Research Laboratory (CRL), and the High Energy Accelerator Research Organization (KEK). International collaborative research was performed by constructing an international broadband network and teaming up with organizations like University College London (UCL) and the European Organization for Nuclear Research (CERN).

2. Research and Development Summaries

2.1 Control Technology

The network control technologies which form the basis for architecture and

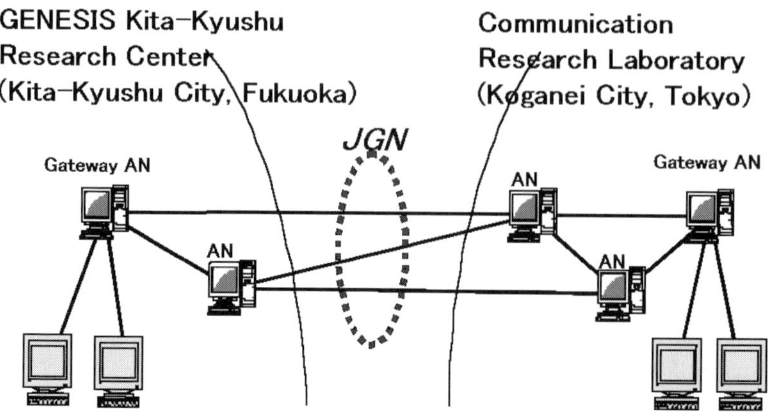

Fig. 2. Active Network testbed.

operation of the next generation Internet were studied from many perspectives under the sub-topics of "Multi-QoS (Quality of Service) network control", "diversity of service quality in a multi-class environment", "IPv6 backbone networks", "network-friendly multicasting", and "applicability of TCP/IP on the gigabit level".

In "Multi-QoS network control", the new paradigm of active networks was researched as a possible control solution for flexibly and adaptively controlling detailed QoS demands from a variety of different users. A stream code active node prototype was deployed on a PC base and its performance evaluated. Also, a broadband experimental architecture (testbed) was constructed on the JGN (Figure 2). Trial versions of active applications (server load balancing and optimum path selection, multicasting with inter-node re-transmission functionality, etc.) to run on active nodes were created and verified in experiments which pointed to the latent potential within active network technology [1] and clarified the challenges that lie ahead for practical application.

In "diversity of service quality in a multi-class environment", Diffserv (Differentiated Services), distributed QoS routing, and concentrated QoS routing were researched as potentially useful technologies for QoS networks. The performance of existing methods was analyzed so that concrete proposals for improvements could be made in terms of allowing QoS traffic and BE (Best-Effort) traffic to coexist efficiently, then numerical and real-world experiments performed to evaluate these proposals. In experiments on Diffserv, methods for the adjustment of RIO (Random Early Detection with In/Out bit) queue parameters for statistical bandwidth allocation service using AF-PHB (Assured Forwarding Per-Hop Behavior) were proposed and verified, demonstrating that high throughput and low delay times were possible. Further, the effects of marking policy type and network topology on Diffserv

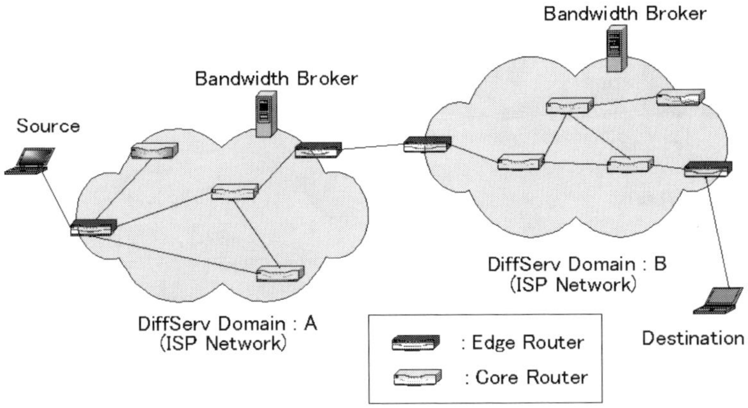

Fig. 3. Diffserv.

QoS for network flow traveling through multiple domains was elucidated [2] (Figure 3). In distributed QoS routing, the OSPF (Open Shortest Path First) routing protocol was expanded and a method proposed for creating routing tables for QoS traffic based on available bandwidth and for BE traffic based on link capacity. This was then demonstrated to result in improved network efficiency. In concentrated QoS routing performed by consolidated QoS (Bandwidth Broker) servers, methods for re-routing traffic according to class and for route allocation taking BE traffic performance into account were proposed and found effective in trials.

In research on "IPv6 backbone networks", high capacity content distribution and high precision time synchronization were studied. For high capacity content distribution, a native IPv6 network was constructed and experiments performed wherein teleconferencing using real time, high capacity transmission was performed to evaluate performance and bring problematic areas to the fore. For high precision time synchronization technology, the first step was to build a prototype PC-based high precision time server, which received external signals from a high precision crystal oscillator or a cesium atomic clock and then used the nanokernel in Linux to realize time precision on the level of a 200 nanosecond standard deviation [3].

In "network-friendly multicasting", real time video streaming (video multicasting), which is expected to be one of the "killer applications" in a broadband network, was studied with an eye to reducing server and network burden while responding to various quality requirements and network conditions from different clients. Proposals were also made for control technology which would coexist efficiently with existing data communications, and numerical and real-world experiments were performed to evaluate these proposals. First, a framework for using an active network

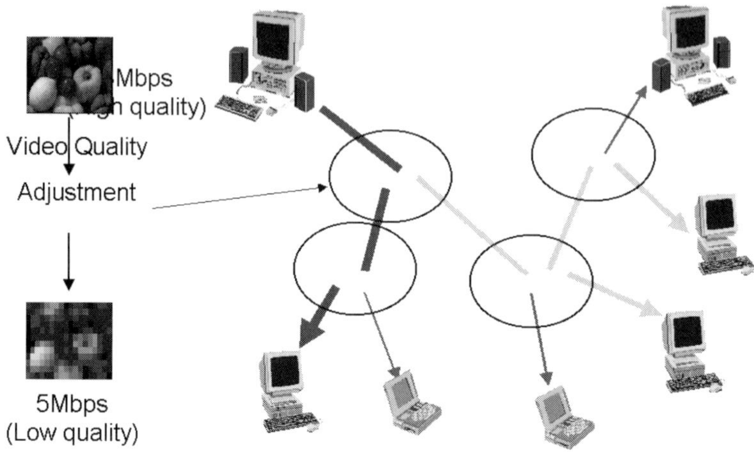

Fig. 4. "Network-friendly" multicasting.

to achieve video multicasting was studied, then elemental technologies such as efficient multicasting group structure algorithms (taking into consideration factors such as network topology and differing quality demands according to clients) and quality control mechanisms performed on active nodes were proposed [4], deployed on the network processors, and proved effective (Figure 4). Also, in order to enable video multicasting taking factors such as competitive traffic and fairness into account, methods were proposed to estimate TCP communication behavior from network load status and use this as a basis for controlling video transmission rates. This system was proved effective with MPEG-2 and MPEG-4.

In studying "applicability of TCP/IP on the gigabit level", data transmission performance using TCP on a broadband network was studied and various proposals made for server-side (end host) technology enabling effective use of bandwidth and fairness of data transmission. After numerical and real-world experiments were performed to evaluate these technologies, each one was found to lead to a significant performance improvement. Specifically, they showed that one of the bottlenecks in the TCP processing of the end host is memory copying, so a proposal for a method reducing the process was made along with other methods for dynamic, fair allocation based on connection properties and status of buffer and other resources required by TCP connections in servers [5]. A method was also proposed for managing flows and local resources on servers such as web proxy servers which conduct data transmission over TCP with both original servers and clients. Such management would result in efficient use of bandwidth without incurring either bandwidth surpluses or shortages. Servers with these new capabilities were used along with existing servers in the same environment in a performance evaluation experiment in which challenges in making the transition to the new technology became clear.

2.2 Application Technology

The application technologies necessary for enabling Internet connections and offering various kinds of services on the next generation Internet were studied under the subtopics of "QoS compliant ISP models and applications" and "an easily-configurable multi-site teleconferencing system". R&D was conducted into architecture for offering realistic QoS services and the high performance applications to run in this environment.

In "QoS compliant ISP models and applications", a practical framework for offering high quality communications services was researched and service models including billing were studied. The necessary QoS server functionality and structure were elucidated and a prototype developed [6]. A user application to interface with this QoS server (teleconferencing using DV over IP) as well as administrator GUI's were also created on a trial basis and their effectiveness demonstrated in experiments (Figure 5).

Multiplayer network games (shared virtual space applications) were studied as a new type of application to run over such a network. Technology was studied for allowing multiple users on different kinds of networks to participate in the same virtual space regardless of the differences in their networks while maintaining certain standards of quality at the application level. The specific game developed was called CittaTron, a network game in which multiple players participated in the same virtual space while

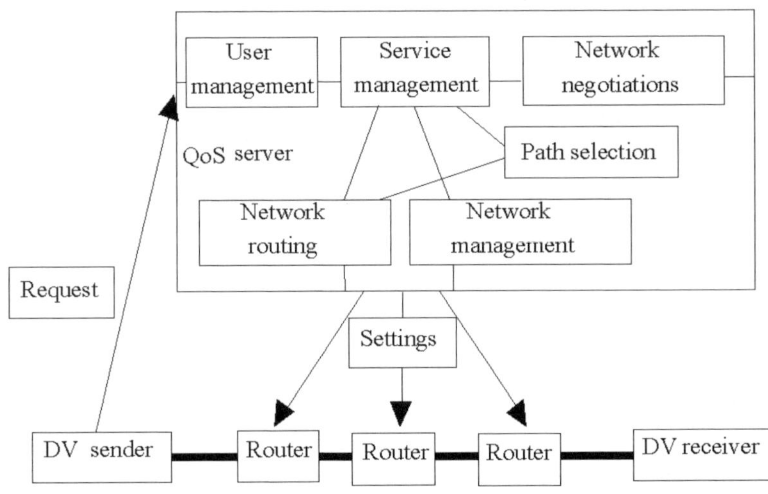

Fig. 5. QoS server architecture.

Fig. 6. CittaTron client GUI.

hunting for treasure. Trial versions of network servers and clients (Figure 6) were created for this game, and methods were proposed [7] for automatic adjustment of elements such as transmission frequency for game movement information and minimum virtual space units depending on the number of players present at any given time, as well as methods for efficient server distribution, client-side server selection, and distributed management among multiple servers. Numerical and real-world experimentation of these elements proved their effectiveness.

In "an easily-configurable multi-site teleconferencing system", high quality multi-site teleconferencing was studied as one of the likely key applications on a broadband network. The object of the research was to resolve challenges standing in the way of wide adoption of such teleconferencing, including such aspects as pre-conference configurations (settings and adjustment of network and audio-visual devices) and various operations during the conference itself (dynamic control of audio-visual devices). Semi-automated technologies which would allow even amateurs in the fields of networking and audio-video devices to use such a system easily were studied. Specifically, a prototype system was developed in which servers were installed at each location to control local audio-visual devices and another server installed to make adjustments for the entire system. These servers initialized each device and performed settings of devices and network connections between locations in a plug and play manner. The system also included functions for using software agents to automatically switch among multiple cameras and views within the teleconference depending on the movement of the participants. The effectiveness of the entire system was demonstrated in experiments.

2.3 Measurement Technology

Communication quality and network status measurements are necessary in order to design, implement, and evaluate the kinds of network controls and applications described above. However, the wide area, broadband, distributed management nature of the next generation Internet can make direct measurement of local and/or global status in networks either problematic or

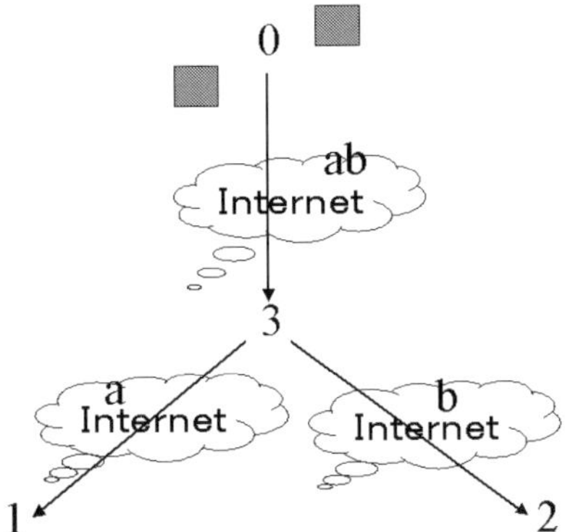

Fig. 7. Packet loss estimation.

inefficient. Thus, technology for indirect measurement of such characteristics through other, more "easily-measurable" characteristics were studied, and methods for enlarging the applicable range of statistical inference using correlations between multi-site, simultaneously measured data (network tomography) were proposed and evaluated through both numerical and real-world experiments.

Specifically, a method was proposed for inferring packet loss rates in individual path segments (identified by the set of paths sharing the segment) by measuring end-to-end packet loss over multiple paths [8]. A prototype inference tool using this method was developed and its effectiveness demonstrated (Figure 7), allowing for localized status within networks to be inferred by measurements performed at end nodes or edge routers. Proposals were also made and demonstrated effective for the statistical inference of arrival rates of individual flows (identified by the set of measurement points through which the flow passes) by measuring traffic intensity (arrival rates) of aggregated flows at multiple measurement points. This method allows for flow characteristics to be inferred without referring to information on origin and destination IP addresses for individual packets.

3. Experimental Network

This project conducted many real-world experiments using international lines and networks such as the JGN as experimental networks (Figure 8). The main networks used are shown in Table 2, and several representative experiments are described below.

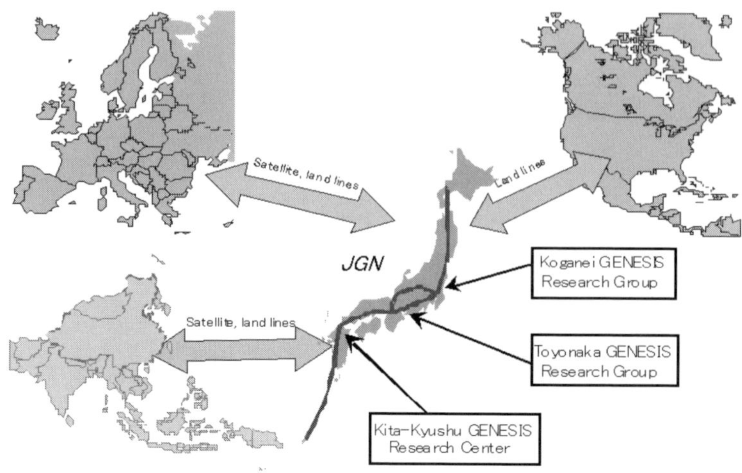

Fig. 8. Wide-area experimental network.

Table 2. Main domestic and international experiments

Era	Experiment
10/97–2/98	Japan–Canada HDTV teleconferencing experiment through GIBN
04/98–	Various remote high-speed communication experiments through JEG
07/00	Multi-site DV over IP experiment (DV Land) at INET2000
10/00	Multi-user network application load balancing experiment
11/00	Active application evaluation experiment
12/00	Wide-area broadband IPv6 network operation experiments
01/01	Multi-site teleconferencing between ICOIN15 in Japan and Europe
10/01	DV live broadcast experiment using IPv6 multicasting over DDW-Japan
11/01	QoS server demo experiment at SC2001

3.1 International Experiments

The GENESIS project participated in the GIBN (Global Interoperability
for Broadband Networks) and JEG (Japan-Europe Gamma) projects, in
which long-distance communications experiments were conducted. One of
the experiments utilizing the JEG project network was the "INET2000 multi-
site DV over IP broadcast experiment (DV Land)" in July of 2000. This
experiment took place at the INET2000 conference held in Yokohama, where
DV over IPv6 was used to transmit video of the Mont Blanc mountains in
Geneva from CERN to the conference center over a 20Mbps Japan-Europe
network. The JGN was also used to transmit live footage from Sapporo,
Kanazawa, Kita-Kyushu, and Naha demonstrating DV over IPv6 technology.

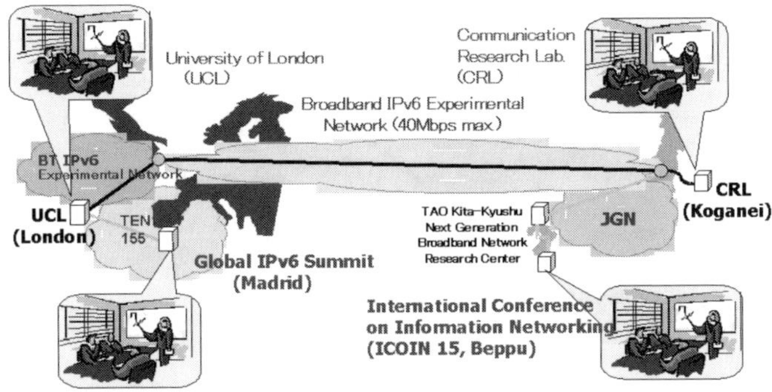

Fig. 9. Japan-Europe multi-site teleconference.

Also, in January of 2001, the GENESIS project teamed up with CRL to construct a permanent IPv6 broadband experimental network spanning Japan and Europe. This network consisted of an extremely long permanent 20Mbps fiber-optic cable stretching from CRL in Koganei through North America to UCL in London. One experiment performed on this network was 4-site teleconferencing using DV over IPv6 at the end of January, 2001, between Koganei (CRL), Beppu (ICOIN15), London (UCL), and Madrid (IPv6 Summit).

Line capacity was increased to 40Mbps for the day of the experiment, as shown in Figure 9. In order to broadcast audio and visual signals to the other three sites, they were first compiled and edited at Koganei before being re-broadcast to each individual site. The only trouble experienced during the experiment was losses of data from Japan to Europe, but the conference itself was not affected by this problem because adjustments were made in retransmission quantity of the video signal. Thus, the experiment succeeded in realizing interactive, multi-site communication on an international level, demonstrating that broadband networks can be utilized for such real time communication.

3.2 Domestic Experiments

The JGN was used on a domestic basis for testing and experimentation related to a variety of research objectives. In research on "QoS compliant ISP models and applications", a wide-area experiment on load balancing functions using multiple servers was carried out in October, 2000, with a multi-participant network application (CittaTron). The experiment took place at three sites (Okinawa, Kurashiki, and Sonobe) connected to each other by the JGN backbone. A server (VSS, or Virtual Space Server) was located at each site to which each client sent its movement information and

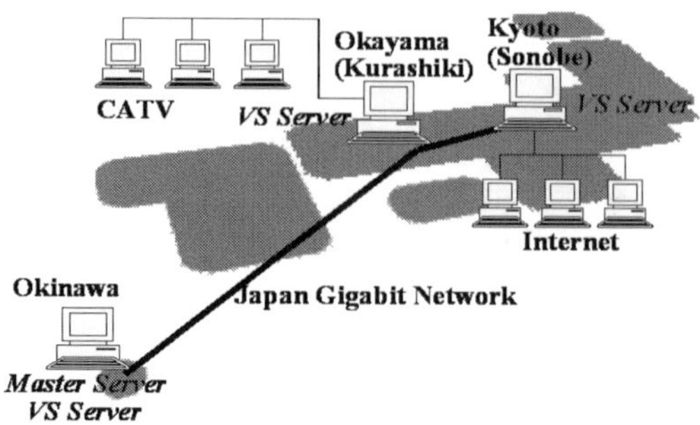

Fig. 10. Multi-user network application game.

from which it received virtual space information. In addition, the master server for managing user log-in and log-out information and pairing clients with VSS's was located at Okinawa (Figure 10). Experiment time and date information was announced on the Internet, the client program distributed, and arbitrary users allowed to participate over the Internet. Clients were also able to participate in the experiment directly over a certain CATV network. In this experiment, when about 50 users connected to the system simultaneously, although the load balancing among VS servers functioned well, a performance bottleneck was found at the master server. However, an additional experiment using an improved version of the master server and two VS servers exhibited better performance with effective load balancing, and verified larger scalability.

In research on "IPv6 backbone networks", an IPv6 native wide-area broadband network for operation experimentation was constructed from December, 2000, in cooperation with CRL. This network was divided into Kanto, Kansai, and Okinawa regions, using high capacity 2.4Gbps fiber-optic networks and gigabit Ethernet with WDM for communication within each region to realize a large scale IPv6 native network. Inter-area connections were accomplished with a 600Mbps connection between the Kanto and Kansai regions and a 124Mbps connection between Kanto and Okinawa over the JGN using IPv6 communications protocols. This experimental network was used on a regular basis for network operation and broadband communication experiments in which issues such as high-speed performance of current IPv6 compliant equipment were brought to the fore.

4. Summary

The GENESIS project conducted multi-faceted research, development, experimentation, and evaluation of the technologies necessary for the architecture and use of next generation broadband networks in order to solve the problems which lie ahead and explore the limits of the technology. The results of these efforts include evaluation of the adaptability of active networks, improvement of quality control methods using Diffserv, elucidation of the challenges in deployment of wide-area, broadband IPv6 networks, greater efficiency in multicast broadcasting, higher performance TCP/IP communication, large scale virtual reality application development, QoS server development, semi-automated multi-site teleconferencing, and statistical inference of network characteristics. This diverse array of results gained from both domestic and international experiments has largely fulfilled the initial goal of the research mission, and has been crafted into many academic journal articles and used for presentations at various international conferences. One of the most significant achievements was the realization of the first ever IPv6 native Japan -Europe broadband communication network.

The fruits of this project amounted to making optimal use of the capabilities of ultra high-speed links to contribute to the architecture and operation of next generation broadband networks which can handle with diverse information quality levels and efficient multi-site broadcasting. At the same time, structural methods for new multimedia applications and Internet services, as well as measurement methods for internal network status, were investigated with great potential for future usc. Also noteworthy is that the research itself was conducted through use of teleconferencing and other electronic information sharing methods, resulting in smooth collaboration between both domestic and international researchers in the experiments which the GENESIS project actively pursued.

As described earlier, this project came to a conclusion in March of this year. However, in order to further develop the project's research successes, remaining challenges as well as new ones will be studied on a continuing basis in the "network architecture research using active network and other technology" project on the same TAO gigabit network.

Finally, although it was not touched on in this chapter, crucial research was also performed in Phase I of the GENESIS project, without which the efforts of Phase II would have been impossible. I would like to express my gratitude to researchers at Osaka University, Kyushu Institute of Technology, Communications Research Laboratory, and the Ministry of Education, Culture, Sports, Sciencc and Technology's High Energy Accelerator Research Organization for their generous support and cooperation in this project.

References

[1] Kubota, F., Egawa, T., Saito, H., Uetsuki, S., Komine, T., Otsuki, H. and Hasegawa, S.: QoS Restoration that Maintains Minimum QoS Requirements – A New Approach for Failure Restoration: IEICE Trans. Communications, Vol. E83-B, No. 12, pp. 2626–2634 (2000).

[2] Kumazoe, K., Hori, Y., Ikenaga, T. and Oie, Y.: Quality of Assured Service through Multiple Diffserv Domains, IEICE Trans. Information and Systems, Vol. E85-D, No. 8, pp. 1226–1232 (2002).

[3] Kitaguchi, Y., Okazawa, H., Shinomiya, S., Nakagawa, S., Kidawara, Y. and Hakozaki, K.: Development of Highly-Accurate Timer Server for Measurement of the Internet, Proc. ICOIN16, Korea, 9C-4.1–4.8 (2002).

[4] Akamine, H., Wakamiya, N. and Miyahara, H.: Heterogeneous Video Multicast in an Active Network, IEICE Trans. Communications, Vol. E85-B, No. 1, pp. 284–289 (2001).

[5] Yagi, H., Manzoor, H., Kitani, M., Baba, K., and Shimojo, S.: Network Functionalities Necessary for QoS Service Provisioning, Proc. IEICE/IPSJ Internet Workhsop 2001, Tokyo, pp. 165–170 (2001).

[6] Hori, M., Iseri, T., Fujikawa, S., Shimojo, S. and Miyahara, H.: CittaTron: A Multiple-Server Networked Game with Load Adjustment Mechanisms on the Internet, Proc. the 2001 SCS Euromedia Conference, Valencia, Spain, pp. 253–260 (2001).

[7] Tsuru, M., Takine, T., and Oie, Y.: Inferring Link Loss rates fro Unicast-based end-to-end Measurement, IEICE Trans. Communications, Vol. E85-B, No. 1, pp. 70–78 (2002).

Chapter 7

Testing of MPLS-Based Wide-Area Distributed IX

Hiroshi Esaki[a], Yutaka Kikuchi[b],
Ken'ichi Nagami[c] and Ikuo Nakagawa[c]
[a] The University of Tokyo;
[b] Kochi University of Technology; [c] Intec NetCore Inc.

1. Introduction

When MPLS (Multi-Protocol Label Switching) [1] is used in IX's (Internet eXchanges), the result is an MPLS-IX architecture [2], a technology which will be described in this chapter along with a report on the wide-area distributed IX currently deployed on the JGN (Japan Gigabit Network).

IX's are autonomously operated structures consisting of multiple interconnected networks. When multiple ISP's (Internet Service Providers) connect to each other, they exchange routing information in a process called "peering", or interconnection [3]. Several hundred IX's exist on the Internet today [4], and they perform the crucial function of exchanging traffic among ISP's. Some world-famous IX's are PAIX, MAE, and LINX, while the largest ones in Japan are NSPIXP2, JPIX, and JPNAP, with other many other regional IX's such as TRIX, OKIX, and TOYAMA-IX serving more local roles.

The authors have pursued research and development of MPLS-IX as a next generation IX architecture. MPLS-IX architecture establishes virtual paths (communication routes) with MPLS between ISP's. This structure allows ISP's to conduct interconnections in an environment independent of the type of data link layer. This same architecture allows control networks which enable virtual paths to function as IP networks, enabling flexible arranging of IX topology and realizing interconnections in wide-area distributed environments.

In the first half of this chapter, we will describe general IX technology and the MPLS-IX architecture proposed by the authors. We will begin with a description of IX structure and touch upon some characteristics of traditional

IX technology. Next, we will detail the enabling technologies of MPLS-IX architecture and explain some of its characteristics as well.

The second half of this chapter will introduce the interconnection testing and experimentation which as been performed with MPLS-IX architecture. The authors founded the Next Generation IX Consortium to pursue R&D of MPLS-IX and conduct tests on its technology. The organization also conducts interconnection tests using MPLS-IX architecture over the JGN network. We will introduce the Next Generation IX Consortium and also report on the MPLS-IX tests which have been performed by this group and others. We will focus on the system and connection structure of the tests performed on the JGN.

2. IX – Internet Exchange

This chapter will discuss IX (Internet eXchange) structure and technology. First, in order to understand some of the characteristics of IX's, we will discuss private peering and IX systems, followed by descriptions of the two major IX technologies in use at present: LAN (Local Area Network) and ATM (Asynchronous Transfer Mode). The characteristics of both of these technologies will be explained here, but for more detail please refer to Ref. [5]).

2.1 Private Peering and IX

When ISP's interconnect, they first form a connection on the physical level and then use BGP4 (Border Gateway Protocol version 4) to exchange routing information and finally data traffic. The different kinds of physical connection which can be used divide all interconnection methods into the two types of "private peering" and" IX".

Private peering entails establishing a dedicated circuit between two interconnecting ISP's. Since this allows for interconnection of ISP's in an independent environment, a high level of freedom in aspects such as physical structure and traffic control is possible. However, this also means that a dedicated circuit must be established for every interconnection between ISP's.

In order to achieve a full-mesh ISP interconnection environment using private peering, the number of dedicated lines necessary is expressed with the formula $N \times (N -)/2$, where N is the number of ISP's. In other words, the number of dedicated lines will reach $O(N^2)$, clearly demonstrating a lack of scalability.

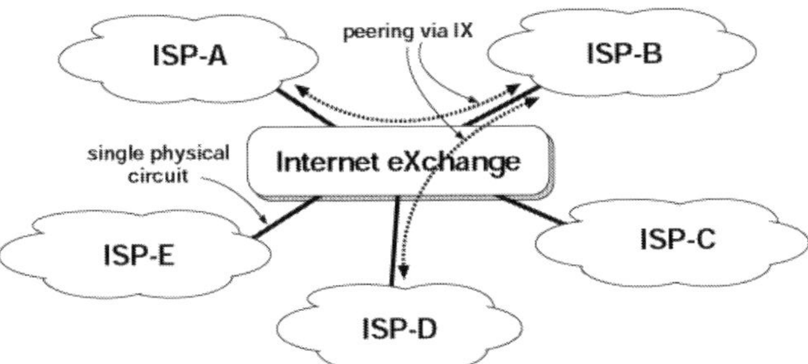

Fig. 1. IX structure.

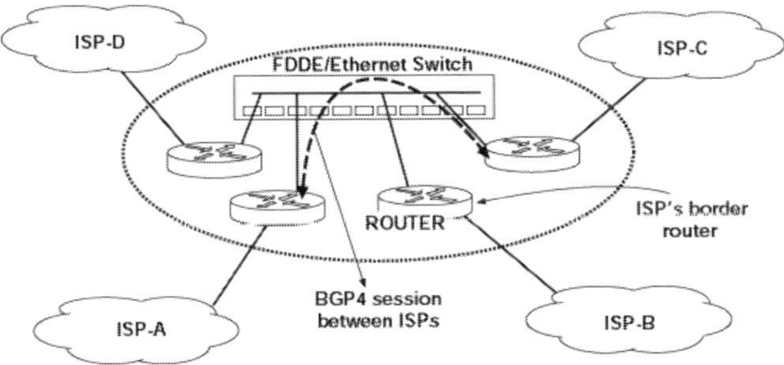

Fig. 2. LAN-IX example.

IX allows for an efficient inter-ISP connectivity environment compared to private peering. It does this by providing an interconnectivity space which allows each ISP to connect their networks to the IX. ISP interconnections may be carried out within the IX, and an environment functionally equivalent to the full-mesh interconnection of private peering is attained (Figure 1). This is the reason that interconnections carried out over IX are sometimes referred to as "public peering".

When using IX for interconnection between ISP's, the number of dedicated lines required is only $O(N)$, illustrating the excellent scalability and efficiency of this approach compared to private peering. From the perspective of individual ISP's, the ability to interconnect with multiple other ISP's with a single physical line allows for traffic consolidation and therefore reduced network costs.

2.2 IX Using LAN Technology

At present, LAN is the most widely-used IX technology. Here we refer to such IX's as "LAN-IX's", the basic structure of which is shown in Figure 2.

ISP's connect to IX's by connecting their individual edge routers with LAN media. Recently, there are many IX's which provide gigabit Ethernet and other LAN switches on the IX side. Each LAN switch is viewed logically as one LAN segment, which functions as a subnet in common for all routers connected to the IX. Each ISP router exchanges routing information on the subnet over BGP4.

2.3 IX Using ATM Technology

There are also many IX's on the Internet utilizing ATM technology. IX's using ATM technology are constructed of either ATM switches or ATM networks consisting of ATM switches. In this case, ISP's connect to IX's through routers with an ATM interface. IX's establish virtual circuits called PVC (Permanent Virtual Circuits) between ISP routers to allow ISP interconnections. Here we refer to such IX's as "ATM-IX's".

In ATM-IX, the PVC's established between each ISP's routers may be treated as point to point virtual circuits. Each router performs routing functions over the virtual circuit using BGP4 while exchanging traffic.

3. MPLS-IX

The authors are proposing a next generation IX architecture which takes advantage of MPLS technology. Here we will describe the MPLS-IX system and it's characteristics [2].

3.1 The MPLS-IX Concept

MPLS is a technology for flexibly routing traffic on existing IP networks by adding a fixed length label to IP packets. MPLS networks are constructed of routers called LSR's (Label Switching Routers), with virtual paths called LSP (Label Switched Paths) established between LSR's to carry data. Labels are attached to packets when data is transported on the network, which are then transported along pre-established LSP's.

In MPLS-IX, interconnections are enabled by LSP's established between boundary routers of ISP's connected to the IX. In general, MPLS's are assumed to have a single management domain; in other words, they are assumed to be used within a single ISP. However, one of the primary features of MPLS-IX is that MPSL functionality can be used among multiple ISP's.

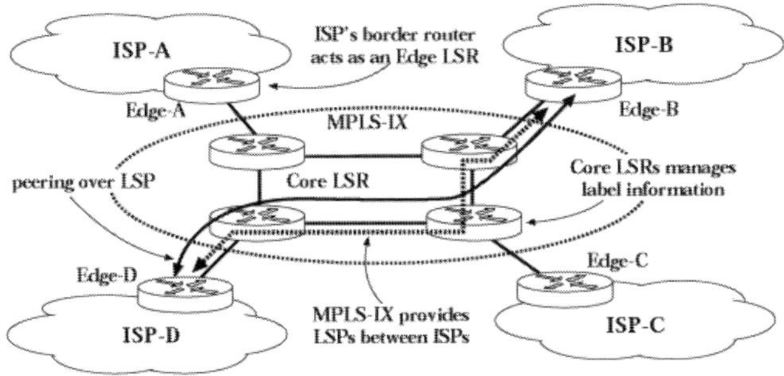

Fig. 3. MPLS-IX schematic diagram.

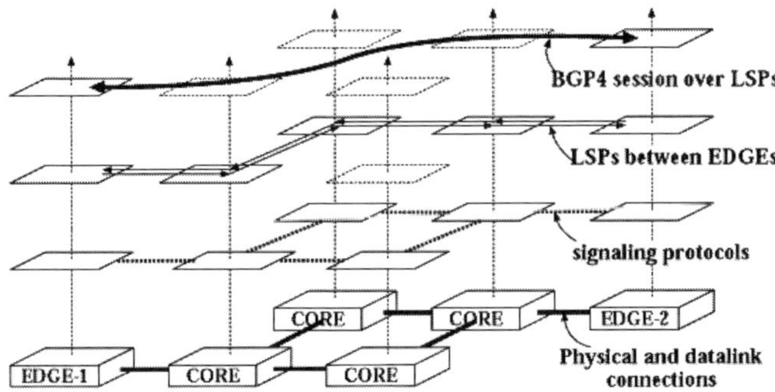

Fig. 4. MPLS-IX architecture.

Figure 3 shows a schematic diagram of MPLS-IX. Individual IX's consist of core LSR's, which are routers controlling the MPLS network. ISP's connected to the IX connect to core LSR's through any edge LSR, which are routers established on the boundaries of the MPLS network. Each edge LSR uses MPLS functionality to establish LSP's between the routers of other ISP's, and then uses this LSP for peering, or interconnections.

Figure 4 illustrates graphically the structure employed for traffic exchange on such a network. As shown in the figure, the MPLS signaling protocol is used between the edge LSR routers of ISP's connected to the IX to establish LSP's. Signaling protocols may be either RSVP-TE (Resource reSerVation Protocol for Traffic Engineering) [6] or LDP (Label Distribution Protocol) [7]. Please refer to Ref. [2] for a more detailed description of the signaling protocol and its operation.

BGP4 is used on the LSP's established by the signaling protocol to exchange routing information. Each ISP then uses this routing information to exchange actual data traffic by transmitting it to other ISP's over LSP.

3.2 MPLS Characteristics

There are two benefits to using MPLS technology for IX. One is the fact that interconnection may be accomplished without depending on a specific type of data link layer. MPLS is designed not to require a specific kind of data link layers or network layer, meaning that data may be exchanged between virtual paths over LSP's even between routers utilizing different data link media. MPLS-IX may perform interconnections over use data link layers such as Ethernet, ATM, or POS (Packet Over Sonet).

In the present Internet environment, the technology used in most IX's depends on either LAN or ATM data link layers, so the ISP's connecting to the IX are not free to choose their data link layer. In contrast, MPLS-IX's may be used with any type of data link layer, permitting data transmission speeds over the IX upwards of 40Gbps by using the fastest technology available, which at present is OC-768 POS.

The second benefit to using MPLS-IX technology is that it allows interconnection in a wide-area distributed environment. MPLS-IX allows normal IP technology to be used in the control network, which consists of connections between core LSR's as well as between core and edge LSR's. Therefore, MPLS-IX allows for a flexible topology, giving the technology superlative expandability by realizing adaptability to wide-area distributed environments, redundant structures, and other network characteristics.

4. Testing of Wide-Area Distributed IX's

The authors founded the Next Generation IX Consortium to pursue testing and research activities of wide-area distributed IX technology using MPLS-IX architecture [8]. The Consortium's goal is twofold: to establish interconnection technology using MPLS-IX architecture and to perform testing of wide-area distributed IX using this technology on the JGN. Here we will summarize the Next Generation IX Consortium's activities and report on the status of its testing program.

4.1 Next Generation IX Consortium

The Next Generation IX Consortium was founded in October, 2001, to conduct practical research into wide-area distributed IX technology using MPLS-IX architecture. Participation in the Consortium is open, and as of November, 2002, it had sixty eight member organizations, most of which are communications services, providers, router venders, universities, and research organizations.

The Consortium currently has three main working groups, each of which pursues research in a defined area.

4.1.1 Router Working Group

In order to realize interoperability utilizing MPLS-IX architecture in a multi-vender environment, the Router Working Group pursues research and testing of MPLS router interoperability. The group has drafted a white paper on the functionality which will be required of MPLS routers in an MPLS-IX. It also provides a forum for exchange of opinions with various MPLS router venders, while also engaging in router interoperability testing experiments in a multi-vender environment with the participation of these venders.

This working group has performed the following interoperability tests, the results for which have been made public.

- First Interoperability Test (10/15/2001–10/19/2001), TAO Tokyo Research and Operation Center. Ten MPLS router venders participated in tests to determine the basic functionality required by MPLS-IX architecture.
- Second Interoperability Test (01/28/2002–02/01/2002), TAO Makuhari Gigabit Research Center. Twelve MPLS router venders participated in tests to determine basic functionality, redundancy, and router performance.
- Third Interoperability Test (05/15/2002–05/17/2002), Netmarks, Inc. Quality Management Center (Tokyo). Twelve MPLS router venders participated in core function testing and performance testing.
- Networld+Interop Tokyo 2002 (07/01/2002–07/05/2002), Makuhari Messe (Chiba). An interoperability showcase for the event was constructed.
- Fourth Interoperability Test (09/02/2002–09/06/2002), TAO Hokuriku IT Open Laboratory (Ishikawa). Ten MPLS router venders participated in redundancy tests and IPv6 interconnectivity tests.

4.1.2 IX User Working Group

The IX User Working Group's mission is to study the functionality and technology requirements for connection from the perspective of ISP's or CSP's (Content Service Providers) connecting to MPLS-IX's, and to pursue testing of such interconnections on the testbed. The JGN was used to construct a wide-area distributed environment as a testbed, to which 20 organizations (as of September, 2002) from all over Japan connected. Testing involved interconnections among multiple ISP's in a wide-area distributed environment, with traffic exchanged to study MPLS-IX functionality. The testing performed by this working group is described in further detail below.

This working group also promotes the sharing of experience and information necessary for MPLS-IX connection settings and operation. The group collects information on settings and operation methods for deployment of a wide array of MPLS routers, and offers this information both within the group itself and over the Internet. Thus, the working group also performs an important educational (human development) function.

4.1.3 IX Provider Working Group

The IX Provider Working Group debates and studies MPLS-IX functionality and technology from the perspective of IX service providers. This group consists of engineers who are involved in designing, constructing, and operating wide-area distributed IX's. Through the experience of constructing and operating the Next Generation IX Consortium's testbed, the members of this group were able to both study and gain valuable experience in the functionality and operating techniques necessary to provide a real-world connection environment using MPLS-IX architecture.

This working group also studied and carried out experiments on a method for interexchange services among multiple providers of an interconnective environment using an MPLS-IX architecture. Specifically, the testbed which the Next Generation IX Consortium was using for experimentation on the JGN was connected to other MPLS-IX providers, and experiments performed on a system allowing ISP's and CSP's to interconnect using multiple MPLS-IX's. Presently (as of September, 2002), Japan Telecom and MCI WorldCom are providing an interconnective environment on an experimental basis on which interexchange services with the Next Generation IX Consortium are performed.

4.2 Experiment Summaries

Following is a description of the testbed constructed and operated by the Next Generation IX Consortium, which is an experimental wide-area IX network with MPLS-IX architecture. As noted earlier, the experimental network is constructed on the JGN, and interconnection experiments are performed mainly with organizations participating in the JGN project "Regional Interconnection Experiment Project".

Figure 5 illustrates the testbed topology. At present, the wide-area distributed IX testbed has six core LSR's, which use Juniper M10's or M20's. These core LSR's are distributed in Tokyo (U-tokyo, Notemachi, Kotemachi), Osaka (Dojima), Toyama, and Kochi, and are connected over the JGN with ATM PVC. The remaining boxes in the graph indicate the names and AS (Autonomous System) numbers of connected ISP's. Dotted boxes indicate planned ISP's. The core LSR's used in this experiment were located at University of Tokyo in Tokyo, the NTT Otemachi Building in

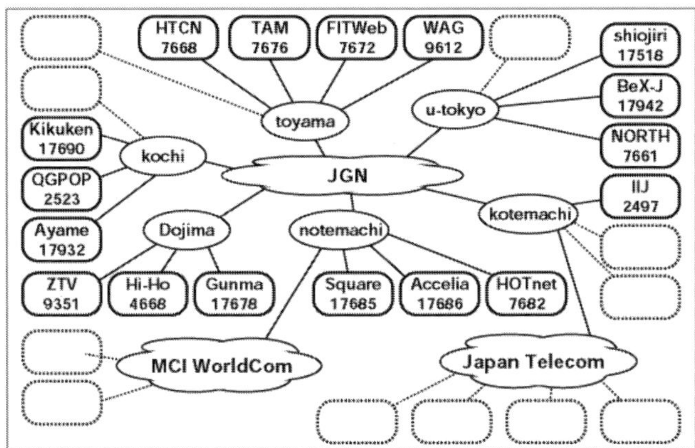

Fig. 5. Testbed topology.

Tokyo, and the KDDI Otemachi Building in Tokyo. This placement was determined because it was judged effective for the purposes of a practical interconnection experiment to use connection points at network nodes where domestic ISP's are already concentrated.

The figure also illustrates interexchange services. Interconnection with other MPLS-IX providers is positioned as one of research activities of the IX Provider Working Group. As described earlier, Japan Telecom and MCI WorldCom are providing an MPLS-IX interconnection environment on an experimental basis, and are performing interexchange services with the Next Generation IX Consortium's testbed.

4.3 Experiment Participation

New groups are always free to join the experiments being conducted on the Next Generation IX Consortium's testbed. When a new organization wishes to participate in an experiment, they must apply to the Consortium officially and connect to one of the core LSR's. There are currently three methods for connecting to one of these core LSR's:

1. Connect over the JGN. The connecting organization connects to the JGN, then establishes an ATM PVC to one of the core LSR's.
2. Direct connection to a core LSR. The connecting organization either establishes their own router or network circuit to make a direct connection to a core LSR. Connection methods such as ATM, POS, or GbE may be utilized.
3. Connect via another provider. Organizations may participate in experiments conducted on the Next Generation IX Consortium's testbed by connecting to an MPLS-IX provider which performs interexchange services.

4.4 Current Status of Experimentation

Currently, 20 organizations are participating in the Next Generation IX Consortium's experiments and exchange of traffic over interconnections. These are a diverse array of organizations including universities, research organizations, ISP's, and CSP's. Since the current focus is on research, most of the experimental traffic on the network consists of communication within specific address spaces, traffic for events, or similar experimental uses.

Figure 6 shows the traffic quantities exchanged over the testbed for these experimental purposes. The graph shows the total amounts of traffic exchanged over the experimental network for the past month in 5 minute increments. As can be seen in the graph, traffic quantities at peak times can exceed the 100Mbps level.

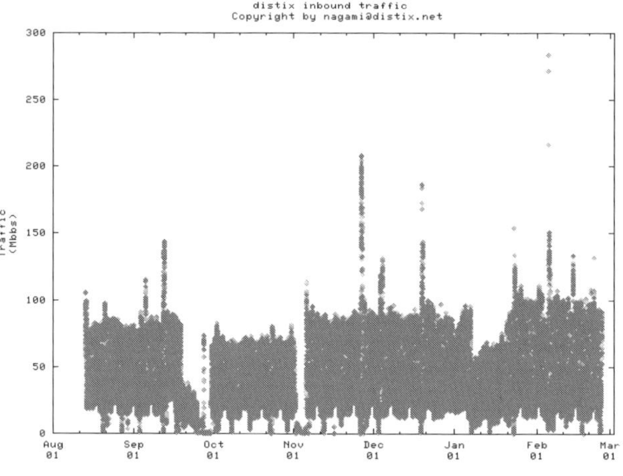

Fig. 6. Testbed traffic.

4.5 MPLS-IX Challenges

Since its inception, the Next Generation IX Consortium has conducted testing and experiments to study the necessary functions of a wide-area distributed IX. In the six months which have passed since interconnectivity testing began, no serious problems have occurred with the interconnections (although there have been planed stoppages). This shows the stability inherent in a system which uses MPLS-IX architecture for wide-area distributed IX.

However, this testing has also brought some challenges into view. Some of these challenges are described below.

1. Since labels are attached to packets in MPLS, caution must be used in setting the network MTU (Maximum Transfer Unit). In general, an MTU

setting between LSR's of 1500 octets or more excluding the MPLS label is recommended.
2. Problems revolving around differences in router deployment remain to be solved. Most MPLS routers already contain the basic functionality necessary for MPLS-IX deployment, but it is impossible to set values – such as BGP4's TTL (Time To Live), for instance – correctly in some routers.
3. It is difficult to detect problems on an MPLS network. Tracing of IP packets on MPLS is difficult, so it can be problematic to identify the location of a problem when one occurs.

5. Summary

In this chapter, we have explained the MPLS-IX architecture technology for wide-area distributed IX's, and given a summary of the experimental network constructed and operated on the JGN. Wide-area distributed IX's have the potential to become the next Internet model. This technology is expected to make great contributions in terms of allowing direct connections between content providers and service providers, inter-regional broadband applications, regional use of major ISP transit services, and regional expansion of commercial IX's.

Meanwhile, tests and experiments carried out by the Next Generation IX Consortium will be transitioning from the first phase, in which basic functionality for wide-area distributed IX's was investigated, to the next phase, in which next generation protocols such as IPv6 and quality control technologies such as QoS (Quality of Service) will be adapted.

Acknowledgements

We would like to thank everyone involved in IX's, but especially Professor Eisukei Hayashi, whose valuable advice helped make this research a possibility. Also, a portion of this research was commissioned by the TAO (121-401).

References

[1] Rosen, E., Viswanathan, A. and Callon, R.: Multiprotocol Label Switching Architecture, IETF Internet-Draft (April, 1999).
[2] Nakagawa, I., Esaki, H. and Nagami, K.: A Next Generation IX Architecture Using MPLS, SAINT2002, Nara (January, 2002).

[3] Huston, G.: Interconnection, Peering and Settlements, The Internet Protocol Journal, Vol. 2, No. 1 (March, 1999).

[4] Manning, B.: Exchange Point Information, http://www. ep.net/

[5] Nakagawa, I., Hayahi, E., Takahashi, T. and Esaki, H.: Technical Trends in Next Generation Internet Exchange, Information Processing, Vol. 42, No. 7 (July 2001).

[6] Awduche, D., Berger, L., Gan, D., Li, T., Srinivasan, V. and Swallow, G.: RSVP-TE: Extensions to RSVP for LSP Tunnels, RFC3209 (December, 2001).

[7] Anderrson, L., Doolan, P., Feldman, N., Fredette, A. and Thomas, B.: LDP Specification, RFC3036 (January, 2001).

[8] Next Generation IX Consortium: http://distix.net/

Chapter 8

Kyushu Gigapop Project

Yuichi Asahara[a], Masaki Hirabaru[b], Motoyuki Ohmori[c],
Koji Okamura[c] and Kenji Watanabe[d]

[a] FM Fukuoka;
[b] Institute of Systems and Information Technologies/Kyushu;
[c] Kyushu University; [d] Saga University

Abstract. The Kyushu Gigapop Project is a research and development project for next generation interconnection architecture and construction of research-oriented infrastructure. It makes use of access line consolidation and aggregation method called Gigapop for use in Internet connections to multiple Internet backbones for research purposes independent of normal Internet traffic. This R&D infrastructure is used for experimental purposes such as online conferences and mobile IP testing. The joint research activities on this testbed have spread not only throughout Japan but also to neighboring countries such as Korea and the rest of the Asian-Pacific region. In this chapter, we give a summary of the Kyushu Gigapop Project.

1. Internet Conditions in Kyushu

When the Internet was just in the process of becoming what it is today in Japan, the Kyushu region saw a spurt of activity in terms of construction and wide use of academic research networks, as well as the formation of Internet communities among researchers and technicians from universities, local governments, and corporations. These communicates engaged in inter-regional cooperation and information exchange, and were instrumental in the construction of one of Japan's first regional networks, the Kyushu Area Regional Research Network (KARRN).

However, after this period the Internet in Japan expanded rapidly. When commercial Internet Service Providers (ISP's) began offering their services on a large scale, the role of such regional networks dwindled, and the unifying force behind the researchers and technicians in Kyushu vanished. However, the great speed at which Internet technology develops has created a new need for a different kind of alliance between Internet technicians and researchers, and for a place where they can meet and exchange information.

It is under these circumstances that the Kyushu Gigapop Project, which conducts R&D for the interconnection architecture necessary for high-speed research backbone connections, was born. Participants in the project include universities, public research institutions, information and communication related corporations, broadcasters, and other kinds of organizations, all of which pursue varied activities in fields such as applied information and communications technologies and digital content creation and distribution.

In this chapter, we will first give a summary of the Kyushu Gigapop Project and present the most fundamental of all our activities, the Gigapop architecture research-oriented Internet. Next, we will introduce a selection of the experiments involved in the Kyushu Gigapop Project. Since we cannot describe all of them within the constraints of this space, we will narrow our focus to teleconferencing, mobile, and streaming experiments carried out jointly with broadcasters. Next, we will introduce the Genkai Project, an international joint collaboration project whose main participants are Japan and Korea. Finally, we will discuss the future prospects of the Kyushu Gigapop Project.

1.1 A Summary of Kyushu Gigapop (QGPOP) and Gigapop Architecture

The growth of the Internet has meant that research institutions now have easy access to a network environment, but this also means that it has become more difficult to make use of the Internet for R&D purposes. In order to pursue R&D on the next generation Internet, it became necessary to establish a research-oriented Internet independent of and parallel to the original Internet. With this purpose in mind, we constructed Kyushu Gigapop (QGPOP) as a research infrastructure utilizing Gigapop architecture which could be put to use for dedicated research into the architecture and R&D environment required for a research-oriented Internet [1].

Figure 1 shows a diagram of the relationship between QGPOP and affiliated research projects. For short-distance connections, Gigabit Ethernet or 100Mbps Ethernet services using dark fiber are used, and for longer distances the Japan Gigabit Network (JGN) is primarily used. External connections with research networks such as the Inter-Ministry Research Information Network or industry networks such as IIJ are performed over BGP, creating an AS (Autonomous System). Connections are also maintained with international research networks in the Asia-Pacific region and North America through APAN (Asia-Pacific Advanced Network) and TransPAC.

Gigapop is an interconnection architecture enabling connection to multiple backbones through a single access point. By using ATM virtual circuits (VC), MPLS, tag-based Ethernet technology (802.1Q), wavelength division multiplexing (WDM), or other technologies to connect to the normal Internet,

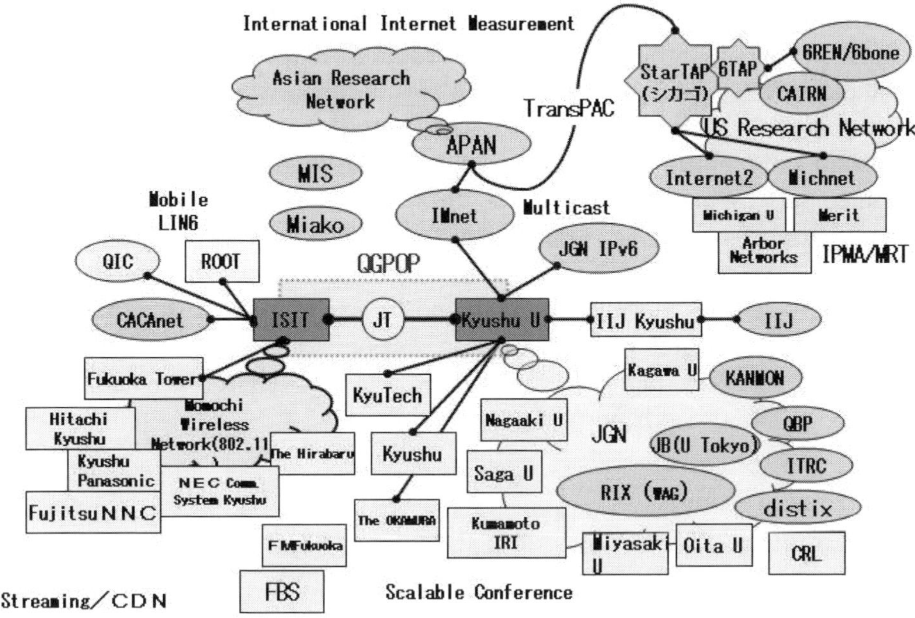

Fig. 1. QGPOP and affiliated projects.

a communication circuit specially reserved for R&D traffic may be accessed over the same physical line. Further, the sharing of physical lines and multiplexing effects have economic benefits as well as bandwidth-saving and other flexibility advantages.

As shown in Figure 2, each organization maintains their own independent internal LAN for research purposes. Fundamentally, these LANs are kept free of firewalls and other impediments to research. Each LAN is connected to the Gigapop over 1 physical and 2 logical access lines, from where traffic may be directed to various R&D, academic research, commercial, or other backbones depending on purpose.

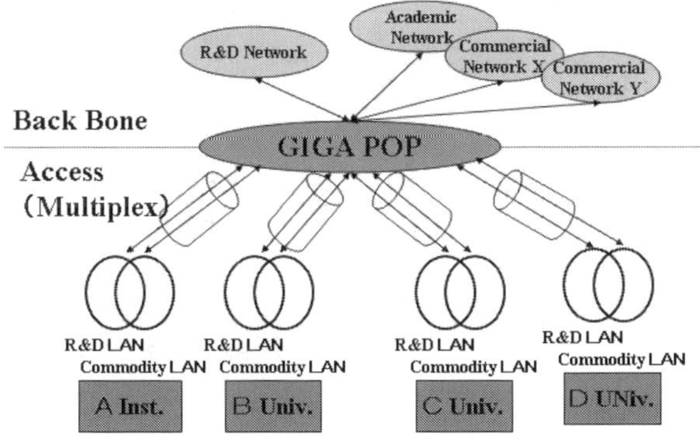

Fig. 2. Gigapop architecture.

Internet usage in research and educational environments has already become common through academic research and commercial networks, but there still remain many research institutions which have difficulties performing R&D on new Internet technologies such as very high-speed broadband communications, multicasting, mobile applications, IPv6, or QoS guaranteeing. If one considers the fact that communications technology is constantly evolving, a high-speed, advanced Internet environment for R&D purposes must be constantly improved and upgraded. Thus, QGPOP effectively combines normal Internet and research-oriented Internet use so that even those research institutions who can only access the normal Internet may access the very high-speed advanced Internet environment constructed on the JGN for research purposes.

2. R&D and Testing

In the next section we will introduce the R&D and testing conducted over QGPOP. Because of space constraints, we will describe only to teleconferencing, mobile, and streaming experiments.

2.1 Teleconferencing Experiments

Two different kinds of meetings are monthly for the QGPOP. One kind is physical meetings, while the other is virtual meeting over the Internet. We call the first "offline meetings" and the second "online meetings". Figure 3

Fig. 3. Online meetings.

shows how the "online meetings" appear to their participants. At present, a user-friendly teleconferencing system based on H.323 is in widespread use, but fundamentally this system is still little more than a toy. However, we are convinced that eventually teleconferencing will become as simple and widespread as e-mail is today, and it is with this practical application in mind that the "online meeting" experiments are performed on a regular basis.

2.2 Mobile Experiments

With the cooperation of Mobile Internet Service (MIS, Inc.), QGPOP conducted an experiment for providing Internet connectivity using MIS's system to participants and others involved in the 9th FINA World Swimming Championships in Fukuoka, 2001. During this experiment, some problems with the MIS system were revealed, but they solved satisfactorily. Presently, the range of this experiment is being expanded around its base in northern Kyushu to test its usefulness. A description of this mobile experiment follows [2].

The purpose of the mobile experiment was to prove the practicality of a mobile Internet environment which moves both physically and in terms of networks while providing high-speed, stable access. Traditionally, such approaches have been limited to cellular and PHS phones, but this low-speed approach requires costs associated with metered billing as well as a lack of IP addresses for i-Mode and EZweb users. Using wireless LAN technology such as IEEE 802.11b solves these problems, but since authentication for each terminal is impossible, a lack of truly secure encryption emerges as a new problem. This is why wireless LAN technology is not suited to use in public spaces such as outdoors or fast-food restaurants.

The MIS system was used in this experiment to solve such issues. Containing elements of both the MIS and Mobile IP Protocols, MIS was developed by Root, Inc., Tokyo Institute of Technology, Kyoto University, Trans New Technology, Inc., and the Institute of Systems & Information Technologies in Kyushu. The MIS Protocol was developed by this group, and solves the problems with wireless LAN while operating in a wireless LAN environment itself. The Mobile IP Protocol, which was developed by the IETF (Internet Engineering Task Force), enables mobile communication over the Internet without breaks.

The MIS system structure is shown in Figure 4. The mobile device referred to in the figure refers to a PC or PDA capable of IP communication. Meanwhile, the MIS Authentication Server is a server which performs authentication service for these mobile devices by managing the private

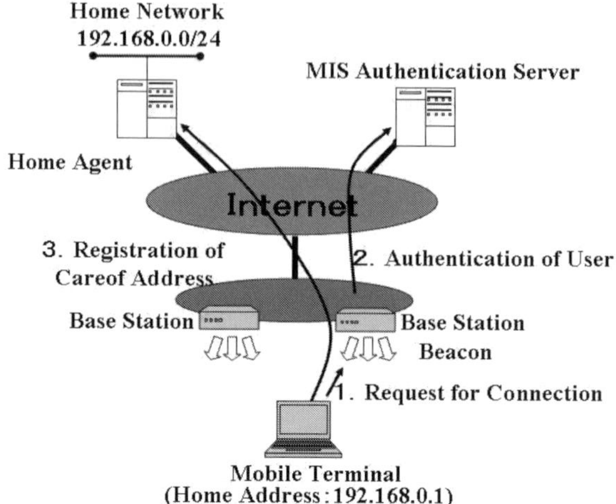

Fig. 4. MIS system.

keys for each device. The Home Agent is a server for mobile IP's. By establishing wireless access points connected to the Internet over land lines in all necessary locations, mobile devices will be able to access the Internet reliably.

The MIS system functions in the following manner. First, the MIS Protocol is carried out. This involves each wireless access point broadcasting packets called "beacons" at regular intervals. Portable devices select which access point to use based on the signal strength of these beacons, then transmit a connection request back to the access point. The MIS Authentication Server then performs authentication by using the private key for that specific mobile device. When this authentication is successfully processed, an IP address unique to that access point called a CoA (Care-of Address) is allocated to the mobile device to enable connection to the Internet. A powerful encryption algorithm called AES (Advanced Encryption Standard) is used to encrypt all traffic between the wireless access point and the mobile device.

Next, the Mobile IP Protocol is carried out. The mobile device possesses an IP address called a Home Addresses which does not change even when it moves, and it uses this address in communications with the Internet host. When the device moves and its CoA changes, the device registers its new CoA with the Home Agent. Then, when Internet hosts transmit packets to the Home Address, they are transferred through the Home Agent to the mobile device using the CoA.

Thus, the MIS Protocol enables communication without breaks and solves the problems inherent in wireless LAN by making use of the Mobile IP Protocol.

2.3 Streaming Experiments

Here we will describe experiments on QGPOP streaming which were conducted with FM Fukuoka and Saga Television Station.

2.3.1 FM Fukuoka Video Streaming Experiment

A video streaming experiment was conducted with FM Fukuoka, which is a participating member in QGPOP, and Hitachi for about six months beginning August, 2001. The structure of this experiment is shown in Figure 5. Servers and encoders used for the experiment both utilized Linux 2.2, with encoders running Real Producer Basic and servers running Real Server Basic. As illustrated in Figure 5, content generated with Real Producer at FM Fukuoka was transmitted to the Real Server at Hitachi over the Internet, from where it was broadcast externally through QGPOP.

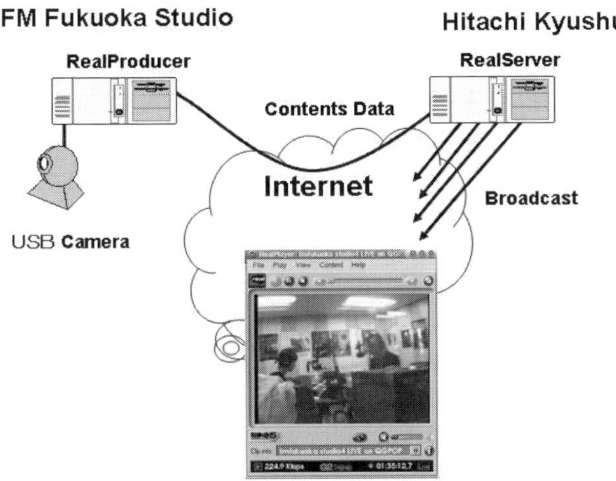

Fig. 5. Streaming experiment.

A Pentium II 500MHz computer with 128MB memory was used as the encoder, but a heavy load on this hardware once the encoding began was observed. In contrast, the server showed no changes at all, even when the number of streaming clients increased. This lead to the conclusion that processing power should be given priority when selecting encoding hardware, but that stability should be given priority for servers. Also, the traffic levels chosen for the encoder's target bandwidth speed were born out in practice, and were observed to match logical values.

The coming age of broadband is expected to entail fundamental changes for broadcasters. Even though this experiment utilized very basic technology, it did provide a concrete measure of what is possible with such technology. Other issues which were revealed during the course of the experiment were peripheral to the technology itself. For instance, copyright protection and income models for copyright holders for content distributed online were among topics identified for future resolution.

2.3.2 Saga Television Station Video Broadcast Experiment

Television broadcasters are experiencing drastic changes brought on by technology, only one of which is the start of digital broadcasting. The issues are particularly serious for regional broadcasters, but these changes also represent a chance to construct a totally new system based on new information and communication technologies. The use of high-speed Internet connections is one of the most important of these challenges. It was against this background that Saga University and the prefecture's only commercial television broadcaster, Saga Television Station, teamed up for a program broadcasting experiment. Here we will explain the purpose and content of the experiment.

Various broadcasting experiments have already been carried out, including live broadcasting of the Koushien high school baseball championships. An environment where teaming up with a "streaming provider" enables easy streaming has already been realized, and Saga Television Station had established a system in advance which allowed streaming for this experiment. The purpose of the experiment was not the broadcasting of the program itself, but rather to 1) perform a "trial run" of such a broadcast that would inform later production efforts, 2) peak interest within both the company and the region, and 3) to provide people who had moved away from Saga with a means to view Saga TV over the Internet. This last purpose, which was the most important motivation behind the experiment, had never before been realized over traditional TV broadcast networks.

QGPOP is used as the network infrastructure for this experiment. Saga University uses the access point established at NetCom Saga to access the JGN, through which it accesses QGPOP. Saga Television Station and Saga University are both located in Saga City, about 1.5km apart from each other as the crow flies. Since the two organizations also have line of site contact, they use a wireless router to connect the encoder server at Saga Television to the broadcast router set up on the QGPOP segment at Saga University. This structure allows for wireless LAN to be used for data transmission alone, while QGPOP's bandwidth can be used for the broadcasting function. NEC's ISS/ES (Internet Streaming Server/Encode Server) and Delivery Server were

Fig. 6. Saga Television Station's video streaming experiment.

used for the encoding server and the broadcasting server to deliver up to 300 streams simultaneously.

The first experiment was conducted from 10/31/2001–11/4/2001. The seasonal events of the "2001 Saga International Balloon Festival" and the "Karatsu Kunchi" festival were material used for a program called "Saga Festivals 2001" which was created for broadcast within Saga. By streaming this program over the Internet as well, the station was able to gain experience for creation of future programs for streaming, research their potential viewership, and generally use the entire experiment as reference for future plans. Some issues which arose during the experiment were copyright of the program music and the status of commercials.

Another requirement for the experiment was stable system operation. This was achieved for the first experiment. Streams were broadcast in low bandwidth (56Kbps) and high bandwidth (300Kbps) versions on Windows Media, but a post-experiment survey revealed that almost all viewers used the high bandwidth stream. The survey also showed that 70% of the viewers accessed the Internet through broadband connections such as fiber-optic, ADSL, or CATV. This result shows clearly the future potential of broadcasting over broadband.

The second experiment is a VOD (Video On Demand) broadcast of Saga Television Station's noon and evening local prefectural news reports. Figure 6 shows a sample broadcast. This experiment's purpose is to establish a procedure for streaming news. Since the news broadcasts are performed on a very strict timetable, they are first automatically recorded to computer. Next the program is edited and encoded into 56Kbps and 300Kbps speeds (both Windows Media) and uploaded to the broadcast server. At the same time, text links are established and the stream advertised on Saga Television Station's website. The news program can be viewed at the following URL. A library system is now under construction. http://www.sagatv.co.jp/

Various programs are under consideration for broadcast this fiscal year, beginning with the balloon festival. Most attention is being focused on the kinds of efforts that best take advantage of the Internet's possibilities. The combination of the Internet with broadcasting has the potential to give rise to a wholly new form of broadcast system not limited by the constraints of traditional media. Saga Television Station's efforts strongly hint at the possibilities.

3. Internationalization of QGPOP – The Genkai Project

The Genkai Project, which is a networking project between Kyushu and Korea, formed as an extension of QGPOP [3–5]. Background for the project is the rapid progress made on the construction of local government high-capacity networks in recent years in Kyushu and Yamaguchi, and the connection of Japan and Korea via an uninterrupted 250km long fiber-optic cable through the Japan-Korea IT Optical Corridor Project. This seabed cable, which began operation in March, 2002, connects Pusan in Korea to Fukuoka and Kita-Kyushu in Japan. The Genkai Project is a large scale, international next generation very high-speed network project which uses the local government networks and this Japan-Korea seabed optical cable. The name of the project comes from the sea through which the cable passes to connect the two countries.

One distinguishing feature of the Genkai Project is that not only research institutions and corporations, but also local governments, are involved. Also, because participation in the project is open, more and more organizations are expected to join, mostly from the Kyushu and Yamaguchi areas. More participants on the Korean side are expected as well.

Discussions on the structure of the Genkai Project's network on both the Japan and Korea sides have been discussed at technical the meetings held concurrently with the six project conferences to be organized thus far. The Japan side is deliberating the international expansion of the Kyushu Gigapop Project from its base in northern Kyushu by using the JGN as a base. Figure 7 shows a diagram of the Genkai Project's structure, focusing on the Japan side.

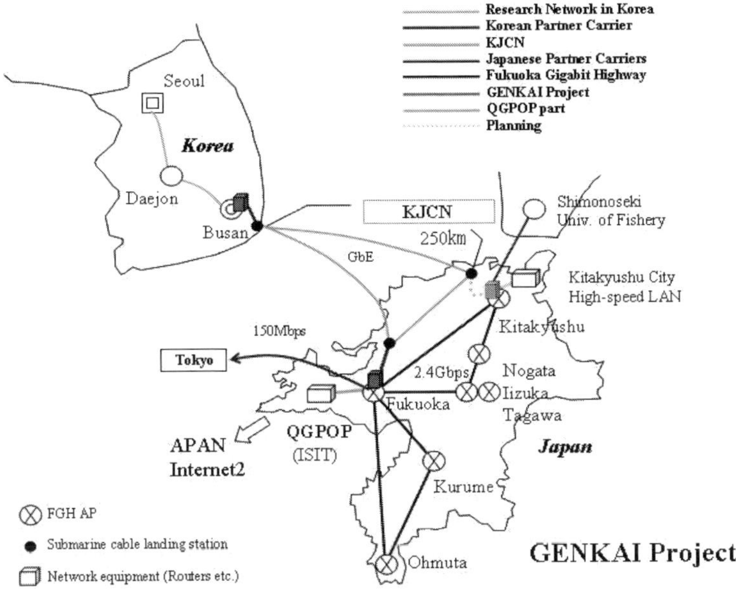

Fig. 7. Diagram of Genkai Project's structure.

As explained previously, the core NOC's (Network Operation Centers) for the Kyushu Gigapop Project are Kyushu University and ISIT, which are connected by a 1Gbps high-speed network circuit and which accommodate all of the traffic from participating organizations. At present, the Kyushu Gigapop Project's upstream networks are IMnet and IIJ, which are connected over a 10Mbps network. By establishing new NOC's at the POI's (Points of Interface) for communication providers established on the KJCN (Korea Japan Cable Network), high-speed interconnections may be performed with the current core NOC's on QGPOP, allowing the Genkai Project's Japan-Korea high-speed network to be treated as one of the upstream Gigapop networks. Plans are also being made to allow the Fukuoka Gigabit Highway and the Kita-Kyushu Regional Information Network, both constructed by local governments, to serve as access points from within Fukuoka prefecture or the city of Kita-Kyushu. All NOC's house high-performance routers which contain IPv6/4 dual protocol stacks compliant with both high-speed layer 2 switches and BGP4+. This allows the experimental devices at each of the participating organizations to make high-speed connections to the backbone as well as accommodate IPv6 networks. Finally, Korea's research network KOREN is being envisioned as the peering partner on the Korean side for the Genkai Project's NOC's on the Japan side.

A high-speed connection between the Genkai Project's Fukuoka NOC and the APAN-JP Tokyo XP in Otemachi, Tokyo, is being considered. Current possibilities being discussed for such a connection are the JGN (operated by the Ministry of Public Management, Home Affairs, Posts and Telecommunications), and SuperSINET (operated by the Ministry of Culture, Sports, Science and Technology), which Kyushu University will connect to in October. If such a high-speed connection with the APAN-JP Tokyo XP can be established, the Genkai Project will be able to function as one part of the Japan-Korea high-speed network from APAN's point of view, allowing higher speed access than is now possible to Internet2 and similar networks in Amerca from the Asia-Pacific region over the Genkai Project and TransPAC. This should have a stimulating effect on researchers in the Asia-Pacific region who would like to pursue their research activities in a truly global environment.

4. Future of the Kyushu Gigapop Project

The Kyushu Gigapop Project began in about May of 2000, and by November of the same year its core had been formed by establishing the main NOC's of Kyushu University, ISIT, and NTT Tenjin. Participation began with national universities and prefectural industrial technical centers connecting over the JGN, and neighboring corporations connecting via the Momochi wireless network. The project experienced many changes in 2002, from the move of the Inter-Ministry Research Information Network, which is one of the project's external access points, from NTT Tenjin to Kyushu University, the removal of the NTT Tenjin NOC, the transformation of the NOC connection between ISIT and Kyushu university to Gigabit Ethernet using a core circuit run by joint research with JT, and gradual but constant changes to the network's topology. The structure of the project will continue to change significantly as it pursues new connections with Korea, Tokyo, and America. However, the status of the Kyushu Gigapop Network as a research-oriented network for researchers will not change. The project will continue to give birth to new technologies and applications which will make important contributions to the world, and it will continue functioning as a nurturer of both network infrastructure and human connections.

Acknowledgements

We would like to express our sincere gratitude to everybody involved in the Kyushu Gigapop Project and the Genkai Project.

References

[1] Okamura, K., Hirabaru, M., Hori, Y., Ikenaga, T. and Arakai, K.: A Summary of the Kyushu Gigapop Project, The Institute of Electronics Information and Communication Engineers (Information Network Research Group), Vol. 100, No. 409, pp. 9–16 (November, 2000).

[2] Ohmori, M., Ohta, M., Nakano, H., Fujikawa, K., Hirabaru, M. and Mano, H.: MIS: The First Public Mobile Internet Service with 802.11b, APAN 2001 Conference in Malaysia (August, 2001).

[3] Okamura, K., Jeahwa, L.: Architecture of the Japan-Korea Joint Research Very High Speed Network, Information Processing Society of Japan, Proc. of Kyushu Branch Symposium (March, 2002).

[4] Araki, K., Okamura, K. and Hirabaru, M.: Korea-Kyushu Gigabit Network: A Frontier Network Research Project beyond the Strain, Proc. of International Workshop on Information and Electrical Engineering (IWIE202) (May, 2002).

[5] Okamura, K.: The Genkai Project, a Next Generation Network Connecting Korea and Japan, IAjapan Review, Vol. 2, No. 2, pp. 2–4 (September, 2002).

Chapter 9

IPv6 Testbed Initiatives in the European Countries

Tim Chown[a] and Jordi Palet[b]

[a] *Department of Electronics and Computer Science,
University of Southampton;* [b] *Consulintel*

Abstract. In this chapter we present an overview of the IPv6 projects being undertaken within the European Commission's Fifth Framework Information Society Technologies (IST) Program and an overview of the IPv6 activities of European National Research and Education Networks (NREN's).

The IST Program covers a broad range of topics across thecomputing and networking spectrum. It includes GEANT, the network connecting the European National Research and Education Networks (NREN's). It also covers projects focusing on IPv6, and those focusing elsewhere but seeking to embrace IPv6 in their scope. The major projects are GEANT, which is now planning introduction of IPv6 services to the existing IPv4 backbone, 6NET, which has deployed a pan-European IPv6-only academic network, and Euro6IX, which is building a network of IPv6 exchange points.

The group of NREN's has an international testbed activity (GTPv6) under which many IPv6 aspects have been studied, including routing, interoperability, DNS, multicast, registries and addressing, firewalls, transition and IPv6 applications.

1. Introduction

The European Commission Fifth Framework Information Society Technologies (IST) Program [14] has a very broad scope, including projects that fall into the area of network research and deployment. Approximately 100 million Euros of funding has to date been directed into projects that are, directly or indirectly, studying IPv6. While this is only a small percentage of the total IST research funding, it represents a significant investment by the Commission. Indeed, because in a typical IST project industrial partners receive only 50% funding (academic partners receive 100%), the total investment is rather higher. While each European country has its own funding methods for academic research, the IST Program is open to collaborative bids from across Europe and thus acts as an excellent stimulus for international, pan-European collaboration, both in terms of research output, and the establishment of facilities such as international testbeds. Non-European partners also cooperate in several projects, e.g. ETRI (Korea) in 6NET, and Russia and China in other cross-Program initiatives. 6INIT was the first IPv6-focused project funded by the European Commission. It began in January 2000 and ran for 16 months. In deploying five national IPv6 clusters with basic IPv6 services, it validated the IPv6 technology, seeding new IST projects. In this chapter, we overview those projects.

2. GEANT

2.1 Overview of GEANT

GEANT (Figure 1) offers a network backbone that interconnects over 25 National Research and Education Networks (NREN's). Available bandwidth varies across the backbone, up to speeds of 10Gbps. GEANT runs on Juniper routers, which have (some) IPv6 functionality in their JUNOS operating system, but this is currently not turned on on the IPv4-only production network. However, GEANT's roadmap has a pilot dual-stack IPv6 deployment due for mid-2003, and a full production service by mid-2004 at the latest. There are already examples of academic networks migrating to dual-stack operation, namely Abilene in the US and SURFnet, Funet and, soon, RENATER in Europe. Thus GEANT's plans may be accelerated. To this end DANTE, who manages the GEANT network, has, with the cooperation of NREN's including RedIRIS (Spain) and ARNES (Slovenia) set a new IPv6 Land Speed Record [18] (with a Tbps result that happens to be higher than the current IPv4 record). There are other aspects to deploying IPv6, including routing protocols, interoperability, DNS services and network management, but performance is a key issue (both for IPv6 and in not adversely affecting IPv4 performance).

GEANT's Task Force for Next Generation Networks (TF-NGN) [17] is undertaking further studies to assist in validation of future IPv6 services. Many of the GEANT NREN's are also in the 6NET project [3].

2.2 The GEANT TF-NGN Working Group

The GEANT Task Force for Next Generation Networks (TF-NGN) [17] meets four times a year, studying a wide variety of topics seen as of immediate or future importance for European NREN's. The participants come from

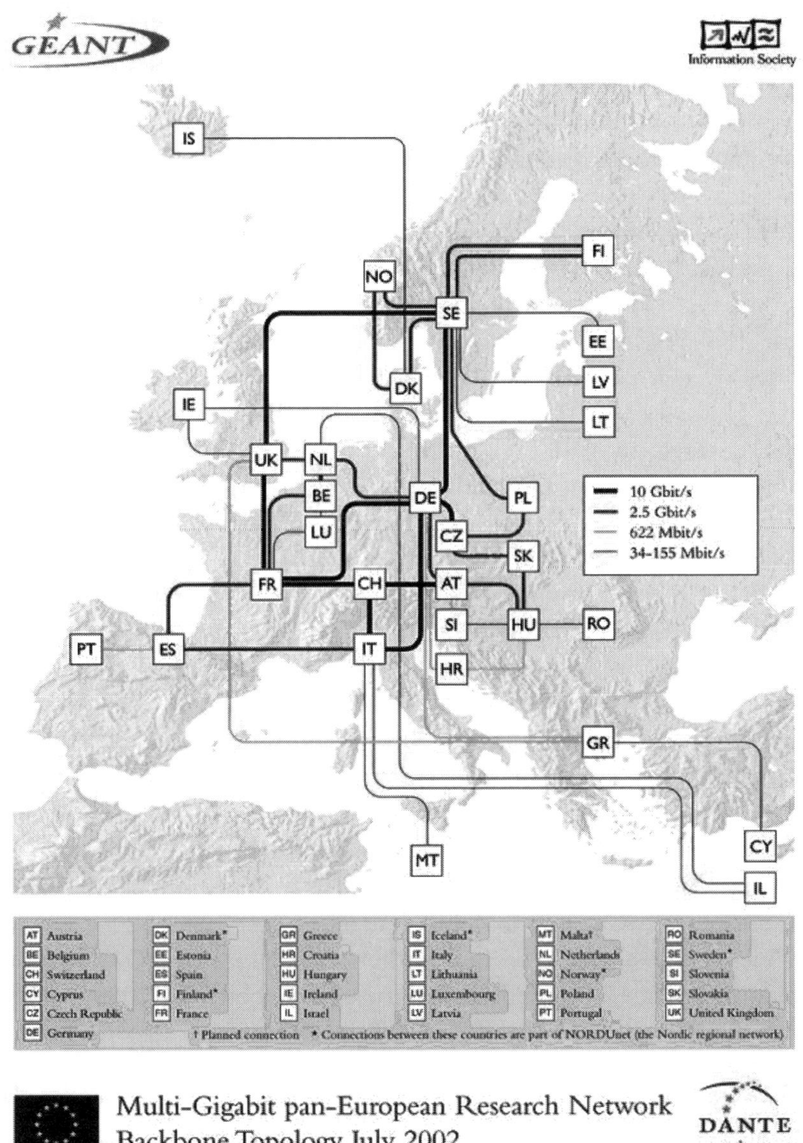

Fig. 1. GEANT topology.

TERENA, DANTE (who manage the GEANT network), the NREN's, and interested universities. The activities include quality of service (including Premium IP and LBE), network monitoring and measurement, IP multicast, optical networking and IPv6.

The IPv6 experiments are carried out within the GEANT Test Program for IPv6 (GTPv6) sub-group of TF-NGN. This group has been active for over four years, and has produced detailed reports of its activities within the GEANT scope in 2001 [24] and 2002 [25].

The goal of the GTPv6 group is to validate IPv6 services with a view to their introduction in the NREN's and on GEANT. While routing performance, reliability and interoperation are key requirements, the successful introduction of (dual-stack) IPv6 services requires a wider consideration of issues such as network management, address allocation policies, peering policies, DNS support, multicast and the provision of transition mechanisms.

Many of the GTPv6 participants are also working on 6NET [23], a three-year IST research project which runs to December 2004. 6NET has already deployed a pan-European native IPv6-only network running at STM-1 speeds. However, the GTPv6 activities continue, working on items not yet being covered in 6NET (e.g. Multicast IPv6), acting as a think tank (e.g. on routing policies) and including involvement from organizations and NREN's not involved in 6NET, e.g. RedIRIS, HEAnet, ARNES, CERN and RESTENA.

The NREN's are generally interested in national and international connectivity. Thus some site-specific issues such as Mobile IPv6, IPsec and firewall provision have received less coverage in GTPv6 than they are receiving in the 6NET project.

1. Juniper M5 Testbed. A Juniper M5 router has been deployed in Paris by RENATER as part of the GTPv6 group's IPv6 experiments [25]. The router has allowed early experience of IPv6 features in JUNOS, and interoperability testing where GTPv6 participants connect to the M5 (receiving /64 prefixes for local experiments). BGP exchanges were set up successfully with Cisco and Zebra routers.

2. Inter-NREN Testbed. In its earlier instantiation, GTPv6 had an international testbed focused on an Ericsson Telebit TBC2000 router, using its own ASN (AS8933) and 6bone pTLA (3ffe:8030::/28) [24]. Since the retirement of the TBC2000, the GTPv6 group is now planning to set up a mini-backbone with the Juniper M5 and a Hitachi GR2000 (based in the UK), using the original ASN and 6bone prefix and allocating /34 prefixes to connecting NREN's and organizations.

   This will enable new interoperability and multicast tests. The original test-bed network was heavily ATM-based, allowing native IPv6 connections via ATM PVC's. Most NREN's have now phased out ATM in moving to higher capacity networks, which means dual-stack routers running IPv6 on the same links as IPv4 is the most likely way to enable native connectivity. The new testbed should be live by December 2002.

3. PC-Based Routers. CESNET has performed experiments with the Zebra PC-based routing platform, comparing Zebra to a Cisco 7500 series router. The results were favorable [25], showing that for an early (parallel) IPv6

deployment, or for site deployment, Zebra is a cost-effective and attractive proposition.

3. Major IPv6 IST Projects

The two flagship IST IPv6 projects are 6NET and Euro6IX [13]. Each has funding of over eight million Euros (more when participant contributions are included), and over 1,100 man months of effort. 6NET focuses on deployment of native IPv6 services in European academic networks, while Euro6IX, with over 1,400 man-months of effort, is deploying pre-commercial IPv6 exchange points for major European telcos. Both projects commenced in January 2002 and run for three years.

3.1 6NET

6NET [3] (Figure 2) includes 34 partners drawn from academia and industry. The bulk of the participants are NREN's and universities, but Cisco (as project coordinators and suppliers of the backbone network equipment), IBM (with interest in e-business, edge services and making WebSphere IPv6-ready) and Sony (for quality of service in multimedia applications) are also project partners.

The 6NET backbone network was specified early in the project. Initially it provides STM-1 capacity links between at least eight PoP's, with new PoP's being added for the three Newly Associated State NREN's recently added to the project (Hungary, the Czech Republic and Poland have been added, an indication of GEANT's requirement to connect Eastern European countries as the EU spreads eastwards).

The main backbone went live in May 2002, with the majority of the PoP's being Cisco GSR routers. University sites generally have 7200-series access routers. Each PoP on the backbone has a corresponding NREN PoP, to which the NREN's own national IPv6 network initiatives are connected. The national projects are not in the 6NET scope, but it is hoped that 6NET will help accelerate the national plans where necessary.

The backbone deployment is only one part of the project (one Work Package). The other Work Packages include IPv6 transition, basic network services (e.g. DNS, Multicast and IPsec), advanced network services (e.g. Mobility, QoS and multihoming), applications (including multimedia and GRID applications), and network management and monitoring. The project has 100 deliverables spread over three years, of which 97 are public, and available from the 6NET website. Of particular interest to date are the

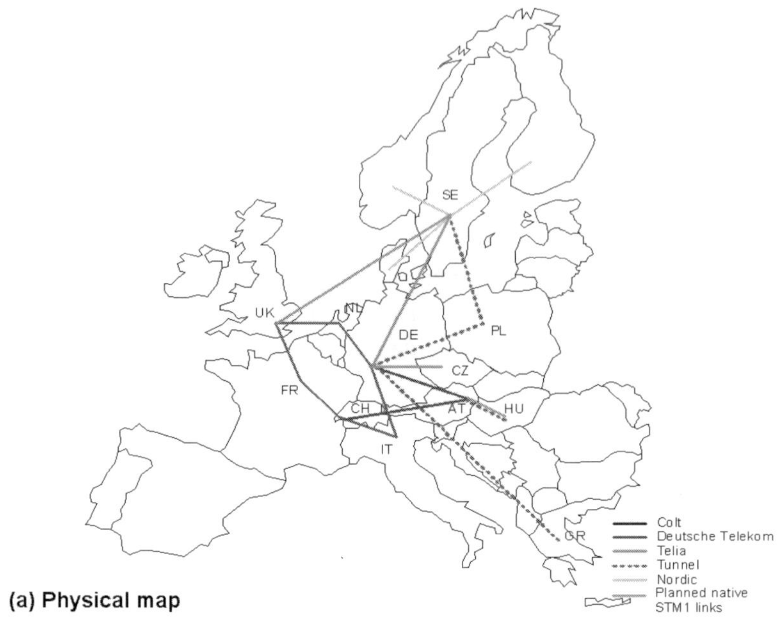

(a) Physical map

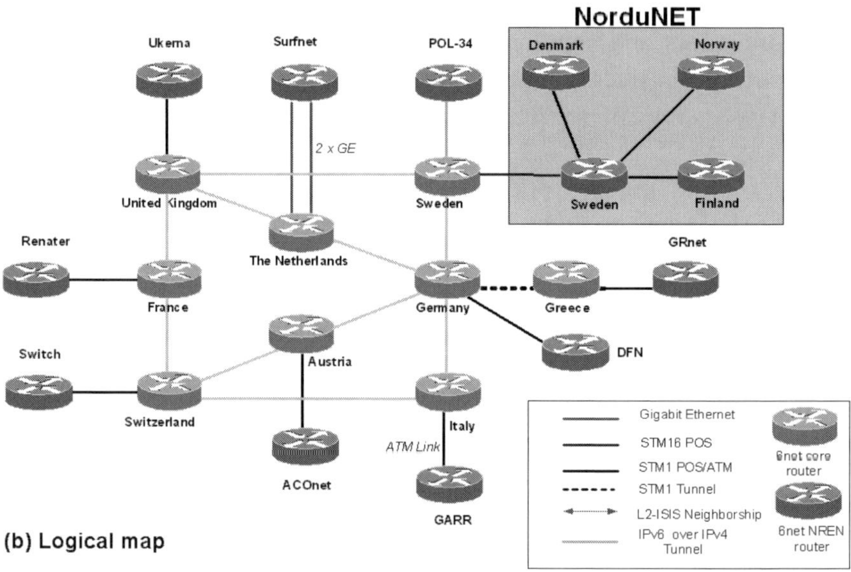

(b) Logical map

Fig. 2. 6NET topology.

transition reports (for university sites [5] and NREN networks [4]), an
evaluation of MIPv6 implementations [6], and a survey of existing and
planned network management and monitoring tools [7]. Other reports include
those detailing the network planning and routing solutions (6NET uses IS-IS
internally), and the DNS structure.

By the beginning of 2003, the project should have produced a number of IPv6 cookbooks designed to assist operators and sites when deploying IPv6 services. There will be cookbooks on topics such as transition tools and network management, including scenarios and configuration examples. The cookbooks will be updated during the lifetime of the project.

3.2 Euro6IX

Euro6IX [13] (Figure 3) is, in contrast to 6NET, focused on telco partners (including those from France, Germany, Italy, Portugal, Spain and UK) and consultancy/systems integrators; of its 17 partners, only three are academic (universities).

The deployment work to date has produced seven IPv6 exchange points, which are located in major European cities [12]. The exchange points are being linked by a 34Mbps backbone (with circuits sponsored by the telcos), to allow experiments in exchange-based addressing and new Layer 3 exchange concepts (to which end the project has obtained a 6bone pTLA). The backbone links have been deployed, and all were operational as of November 2002.

The goal is to investigate new business models for IPv6 deployment, where telcos can come together at exchange points for their own mutual benefit, while offering novel, advanced services to their customers. The exchanges are open for non-project ISPs to interconnect for pre-commercial trials.

The router equipment being used in Euro6IX is more varied than 6NET, where Cisco equipment dominates. Hitachi has provided project sponsorship, with the result that a number of GR2000 series routers have already been deployed. The telcos will also consider IPv6 access technologies. The

Fig. 3. Euro6IX topology.

Euro6IX work items are split into basic service deployment (e.g. DNS), advanced services (including Mobile IPv6, IPv6 Multicast and QoS) and application development and porting. An early deliverable, available from the web site, specifies the applications that will be developed and deployed. These include new multimedia and conferencing applications, plus two partners working on novel messaging systems.

3.3 Common work items

The 6NET and Euro6IX projects have agreed common work items upon which they will collaborate in their three-year lifetime, including Multicast and IPsec. The networks have been interconnected via the UK6X in London, and soon also via Torino and later probably in Paris. Applications and network management and monitoring tools will also be exchanged. The applications and tools will generally be available in open source format.

An interesting collaboration area is defining best practice for application porting, such that applications can run IP independently. Some guidance exists already (e.g. [20]); this has been used as input on the projects. The Vocal VoIP and Globus Toolkit packages are among those being ported to support IPv6.

4. Other IPv6 IST Projects

4.1 6INIT: The First IPv6 Project

6INIT [1] was the first IPv6-specific IST project. In deploying five national mini-clusters it paved the way for the IPv6 projects that followed. ATM PVCs and IPv6-in- IPv4 tunnels were used to link the clusters. Each cluster included deployment of certain basic services, including Web servers, email gateways and DNS servers. The NAT-PT transition tool was used (specifically the Ultima tool from BT Exact) to offer access to external IPv4 sites from the IPv6-only 6INIT sites.

4.2 6WINIT: IPv6 in Clinical Sites

The 6WINIT project [10] concludes in January 2003. Its unique and successful concept has been the marriage of three clinical sites with commercial and academic partners. The main project scenarios were threefold: communication between ambulances and hospitals using IPv6 over UMTS, use of mobile devices in hospital environments, and remote access to patient information. In each case, security and mobility were key requirements. While the UMTS aspect of the project has been delayed for external reasons, the reminder of the project has been very successful. A variety of tools have been demonstrated, from road warrior VPN's to Mobile IPv6 PDA's and advanced media gateway devices including RTP and Java components.

4.3 6POWER: IPv6, QoS and Multicast over Power Lines

6POWER [8] is a new 24-month IST project whose goal is demonstrating that IPv6 over PLC (power line communications) can offer a broadband solution for delivery of IPv6 to end users that is both affordable to the customer and profitable for the supplier. The technology used in the project will enable broadband access initially at 45Mbps, rising to 200Mbps, and will include support for advanced QoS features and IPv6 Multicast. The project includes development of accompanying head ends, home gateways and set-top boxes, and the provision of next generation IPv6 applications such as VoIP. Different Power Line networks across Europe will be connected via Euro6IX and 6NET for trials and demonstrations, for example for the next Madrid IPv6 Global IPv6 Summit (http://www.ipv6-es.com).

4.4 6QM: IPv6 QoS Measurement

The IPv6 QoS Measurement project [9] is also a new 24-month project. Its goal is to develop an accurate IPv6 QoS measurement device using GPS or an equivalent technology. The measurement device is designed to be deployed by operators and ISP's in support of QoS services for end users. It is planned that the 6NET and Euro6IX projects can conduct trials for the device in their own networks. These projects will also deploy their own measurement and monitoring devices (e.g. the IPv6-enabled RIPE Test Traffic Server), which can be used to assist in the development of the 6QM device. A worldwide Measurement trial is expected during the next Madrid Global IPv6 Summit, together with the ETSI IPv6 plug-test.

4.5 Eurov6: The European IPv6 Showcase

The Eurov6 project [15] is building an IPv6 applications and services showcase, bringing together European systems developers, operators, ISP's, application developers, and vendors. Its aim is to make high-profile demonstrations of the business and technology opportunities that exist with the deployment of IPv6. The work mirrors to some extent that of the Japanese IPv6 Promotion Council's Galleria showcase, so several showrooms will be setup across Europe (initially in Madrid, Brussels and Basel), and a Nomadic version will be ready to be deployed in related events.

4.6 LONG: IPv6 Collaborative Working

The most publicized output of the Laboratories over Next Generation Networks [21] project was the IPv6 port of the ISABEL collaborative working (CSCW) system. The project also included studies of IPv6 transition tools for new access methods and advanced network services. ISABEL is now being marketed by Agora, although educational use remains free. ISABEL has been used at many IPv6 conferences to bring remote participants to workshop and presentation sessions. The 6NET and Euro6IX projects both plan to use ISABEL for project conferences.

4.7 SATIP6: IPv6 over Satellite

The Satellite Broadband Multimedia System for IPv6 Access [22] project is a 24-month IST project focusing on the delivery of advanced network services by satellite. It is developing technology required for the interaction between the physical media and the IPv6 layer, migrating existing IPv4 technology where it exists, including support for a variety of interworking and gateway devices. DVB-RCS access will be adapted from existing broadband solutions.

5. IPv6 Clustering

The IST IPv6 Cluster [19] is an important activity for IST projects that include IPv6 activities in its roadmap (over 90 MEuros funding has been provided on this topic in FP5). The Commission encourages and assists project clustering in a variety of themes, including IPv6, Systems Beyond 3G and Grids. The IPv6 Cluster enables IST projects to meet and discuss all aspects of IPv6 research and deployment, both via email and several annual meetings.

The 6LINK [3] project supports the IPv6 Cluster, providing logistical assistance, but also producing documents such as its quarterly newsletter and standardization (principally IETF) reports. 6LINK has also led a series of

All-IPv6-World meetings in which European researchers have been invited to contribute their views on IPv6 research and deployment issues. The meetings have led to a streamlined set of study areas that may lead to new research projects within the Sixth Framework (see below).

6. The EU IPv6 Task Force

The EU IPv6 Task Force [11] began its work in mid-2001. Its aim was to make recommendations towards deployment of IPv6 for the European Commission for industry and for governments. In January 2002, the final report of the Task Force was taken forward to the Spring Meeting of the European Heads of State, out of which arose a directive recommending the deployment of both IPv6 and broadband services in Europe by 2005. The recommendations are now being driven at the national level in the second phase of the Task Force, with the support of a focused, IST-funded Steering Committee project. Since the IPv6 technology adoption paths and processes vary from country to country, the use of national task forces in combination with a continuing EU IPv6 Task Force is considered to be the most productive way forward.

7. The IST Sixth Framework Program

The Sixth Framework Program (FP6) for IST is due to begin shortly. The first call for projects was due in December 2002, with successful projects likely to start by October 2003. In this Framework the funding available for IPv6 projects will come in two main forms: Integrated Projects and Networks of Excellence. The former caters for consortia doing significant research and development work, the latter supports (virtual) centers of excellence.

8. NREN Projects

In recent months, two specific projects of interest have arisen from the NREN activities.

8.1 The m6bone

The m6bone [10,6] (Figure 4) is a Multicast IPv6 test-bed network spanning many European countries and reaching into Africa. The aim of the project is

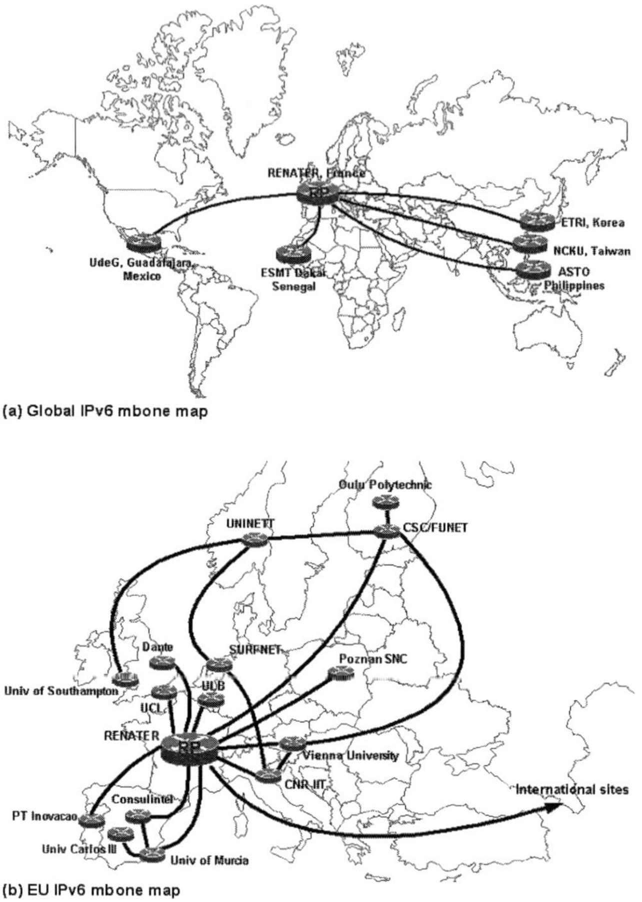

Fig. 4. mbone topology.

to offer an IPv6 multicast service to interested sites. With no pan-European IPv6 network supporting PIM-SM Multicast natively, the m6bone network is almost entirely tunnelled, either IPv6-in-IPv4 or IPv6-in-IPv6. A single RP lies on the PIM-SM BSD router in Paris.

To date, participants have used (Free)BSD (with the KAME stack), 6WINDGate, Hitachi GR2000 and Cisco 7200-series routers to connect to the testbed. PIM-SM is used for the multicast routing, and RIPng for the unicast routing. FreeBSD, Linux and Windows 2000 clients have been used to run the vic and rat multicast conferencing applications over the network. While native multicast is lacking, the testbed has been very useful in diagnosing problems in host and router implementations, and in gaining experience of the tools in an IPv6 environment. Future planned work includes investigation of multiple-RP PIM-SM networks, and use of PIMSSM (as implementations and MLDv2 become available).

8.2 IPv6 Land Speed Record

The Internet2 Land Speed Record [16] was originally devised to promote IPv4 performance improvements, but has recently been extended to IPv6 for single and multi-stream traffic.

As part of the effort to show the viability of running dual-stack IPv4 and IPv6 on the GEANT Juniper routers, DANTE, together with RedIRIS and ARNES (Slovenia), made an attempt to set a new Land Speed Record by running IPv6 on the production IPv4 Juniper (M20, M40 and M160) routers, running JUNOS 5.3R2.4 and R3.3. IPv6 was enabled on the routers, and static routes used, with one path within GEANT via Austria, Switzerland and Italy, and one longer path adding the UK, New York and France.

The attempt used 1.1GHz and 1.3GHz dual Pentium-III PC with Intel Pro 1000/XT and 82543 Gigabit Ethernet adaptors, running Linux 2.4.18 and 2.4.20 kernels with RAM disks and iperf-1.62, wu-ftpd-2.6.2 and ncftp-3.1.3 software. Some TCP parameter tuning was used. Initially, a single-stream record was accepted on 4th October 2002 at 1,215 Tbps; this was then improved with the attempt via New York on 23rd October 2002 at 350Mb/s, which equates to 5,154 Tbps. This is actually some 5% faster than the existing IPv4 single stream record, although of course a direct protocol comparison cannot be drawn from this result. However, it showed that IPv6 forwarding on GEANT was comparable to that obtained with IPv4.

9. Individual NREN Activities

Here we describe individual initiatives by NREN's, the two particularly interesting case studies being in SURFnet and Funet, where IPv6 dual-stack networking has been deployed in the production network.

9.1 SURFnet (Netherlands)

The SURFnet5 network (Figure 5), which was enabled within the national Dutch project GigaPort, was the first broadband research network to provide native support for IPv6. Prior to building SURFnet5, an OC48 testbed was set up with four Cisco 12416 routers running the IOS dual stack image available at that time. At the start of 2001 two core locations in Amsterdam (each with two GSR+ routers) were deployed. Fifteen concentrator locations (GSR+ routers) each connected by OC192 interfaces to both core locations. Each of the concentrator locations has GE interfaces to connect customers. Linux systems with GE interfaces were used to generate IPv4 and IPv6

Fig. 5. SURFnet.

traffic to check that IPv6 load on the network would not adversely affect IPv4 performance.

SURFnet also has a number of relatively low speed leased line customers. To connect those customers to the new network, a Cisco 7507 router was installed with a GEIP+ interface connected to the GSR+ routers. Unfortunately the 12.2T train (needed for IPv6 for this platform) did not have support for the GEIP+ interface, so the plan for a complete dual-stack network was delayed.

During the latter part of 2001 all customers were migrated to the new SURFnet5 network, which was partly dual-stack at that time. Customers could have a native IPv6 connection or a tunnel to the nearest GSR+. There were no problems experienced apart from failed attempts to move from RIP to IS-IS. It is hoped that IS-IS can be used soon (6NET is running IS-IS on Cisco GSRs').

Early in 2002, an IPv6 image became available for the Cisco 7507 which supported the GEIP+ interface. After a successful test, this image was deployed on all Cisco 7507 routers in SURFnet5, resulting in an almost completely dual-stack network (the out-of-band network is currently IPv4 only). Customers connected to a Cisco 7507 and an IPv6-in-IPv4 tunnel to the GSR+ were migrated to a native connection or a tunnel on the Cisco 7507.

Some features are still required. The lack of extended ACL's for IPv6 has sometimes prevented IPv6 connectivity to SURFnet service 'AN's. Measuring the amount of IPv6 traffic flowing through the network can also be difficult, but solutions for these issues will become available. SURFnet has been able to supply enough IPv6 performance for its customers using a dual-stack strategy. The question is whether this will be true if the amount of IPv6 traffic grows, because a large number of interfaces do not (yet) have support for IPv6 in hardware.

9.2 Funet (Finland)

In 2001, Funet's core network (Figure 6) was upgraded to 2.5 Gbps POS links. Six Juniper routers (M20) were added to the previously all-Cisco network. By the end of 2001 plans for enabling dual-stack on the Juniper routers had been made. Tests early in 2002 led to a cycle of problem fixing and further tests, and in Q2/2002 the minimum working version (JUNOS 5.2R2) was upgraded on the routers, followed by versions 5.3R2 and 5.3R3. JUNOS 5.4R2 is due to be tried next as it should fix all the observed problems.

The IPv4 network uses OSPFv2 and BGP for routing, but OSPFv3 for IPv6 was not ready. There was no desire to change the routing protocol, and for clarity separate IGP's for IPv4 and IPv6 were wanted: OSPF and IS-IS. In addition, if only IS-IS was used, multiple topologies would have to have been supported as IPv6 routing would have been different (in parts) from IPv4. Thus IS-IS for IPv6 only was the best (and only, discounting RIPng)

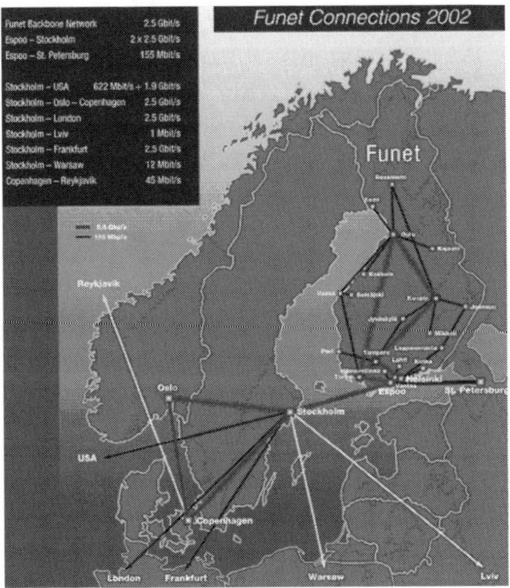

Fig. 6. Funet.

choice. Unfortunately, there were problems in JUNOS and IOS with IPv6-only IS-IS. Juniper would always advertise IPv4 addresses used in loopbacks and point-to-point links. Cisco, unless was enabled IS-IS for IPv4 as well, would discard all such attempts to form adjacencies: a total interoperability problem. Cisco's IS-IS implementation has an option "adjacency-check" to override this; however it would only work using level-2-only IS-IS circuit-type. A first step in interoperability was gained when these were enabled in Cisco.

Further, IS-IS route advertisements from Ciscos to Junipers were accepted in the route database, but not put to Junipers routing table: this was caused by the above problem with adjacencies, and fixed in 5.2R2. The issue with Juniper always also advertising IPv4 addresses in IS-IS was fixed with the "no-ipv4-routing" feature in 5.4R1. Also, one could not redistribute static discard routes to IS-IS (to generate a default route) until JUNOS 5.3R3. Also in 2002, the rest of the Cisco routers were replaced by Juniper M10s and M20s, in part for their IPv6 capabilities but mainly due to scheduled upgrading and phasing out of ATM.

Funet has almost all the IPv6 features it needs at present; only IPv6 multicast is missing. The dual-stack access line is an approach that should be supported in the long term. The next question is how to offer IPv6 to universities without forcing an upgrade of its edge router to dual-stack IPv4/IPv6 (which may not yet be desirable depending on the hardware).

9.3 Renater (France)

Renater (Figure 7) is planning to introduce dual-stack service with its imminent deployment of Renater 3; this could thus be the third European

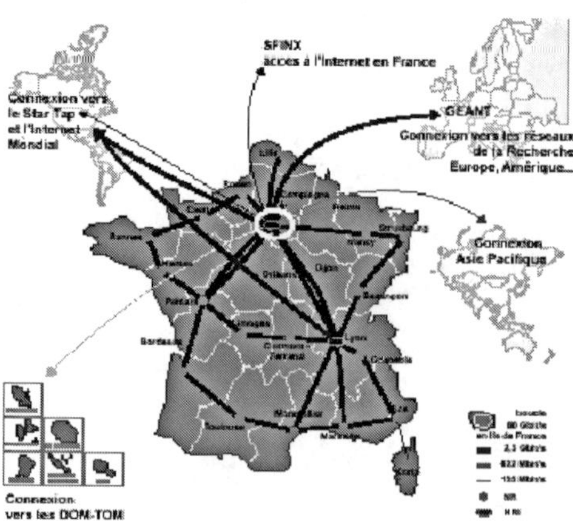

Fig. 7. RENATER.

NREN to move to dual-stack operation. Renater is also leader of the m6bone project described above, and Paris is the European end-point of the Trans-Eurasia Information Network (TEIN), which connects Europe directly to Asia (Korea), currently at 20Mbps via ATM, but soon to be raised.

9.4 DFN/JOIN (Germany)

The JOIN project within the DFN (Figure 8) is notable for having the largest number of sites connected to any European national pilot (over 100). The JOIN network is described in a 6NET deliverable [23].

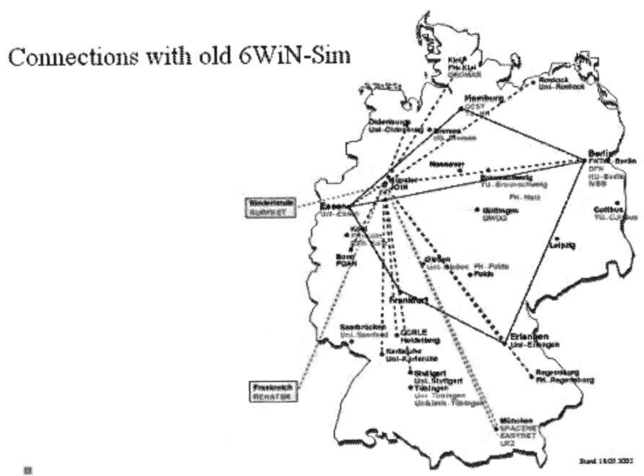

Fig. 8. DFN 6WiN.

9.5 UKERNA (UK)

Since 1999, UKERNA has supported early IPv6 trials, in particular providing management for the Bermuda 2 IPv6 project [23], which utilized native IPv6 links over the JANET ATM Managed Bandwidth Service. However, the new 10Gbit/s POS network no longer supports ATM, and connectivity has dropped back to IPv6-in-IPv4 tunnels for the time being.

In May 2002, UKERNA formally launched a (tunneled) IPv6 Experimental Service on JANET (Figure 9), to which UK Universities and Colleges can connect. The aim of the experimental service is to allow UKERNA and JANET Connected Organizations to gain early experience in operating IPv6 based networks and services, and to focus the JANET community's IPv6 efforts. UKERNA are currently investigating the possibility of implementing dual-stack on the production network during 2003.

Fig. 9. UKERNA JANET.

9.6 RedIRIS (Spain)

RedIRIS (Figure 10) has promoted the use of IPv6 in the ESPANIX (Spanish Internet Exchange Point), and has started to exchange IPv6 traffic with Telefnica, BT Spain and Intelideas. One area of interest has been ongoing DNS testing, where interoperability has been tested between servers with Bind 8, Bind 9, dual stack, IPv6-only and IPv4-only.

9.7 CESNET (Czech Republic)

The CESNET IPv6 network (Figure 11) currently consists of eight nodes connected by configured tunnels which closely mirror the physical topology of the underlying IPv4 network. CESNET is promoting Intel PC's with Linux

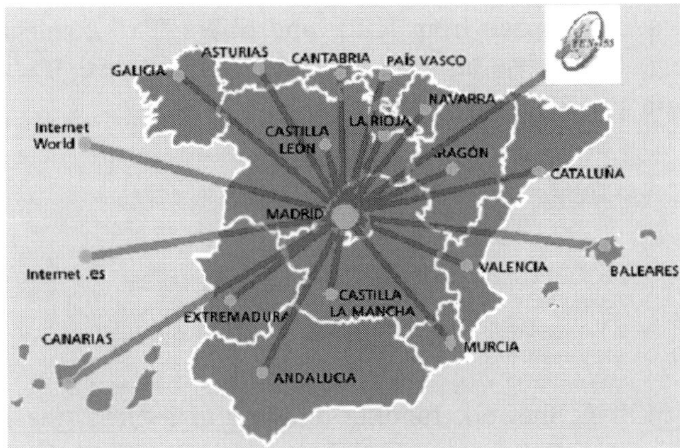

Fig. 10. RedIRIS.

or NetBSD as a general routing platform for IPv6. Therefore, five of the eight backbone routers are such PC's running the Zebra daemons, the rest being Cisco routers. RIPng is used as the interior gateway protocol because it is the only one supported by both Zebra and Cisco IOS. In order to increase the performance of PC routers, CESNET is developing a four-port hardware accelerator (PCI board) aimed at throughput up to 10Gbps.

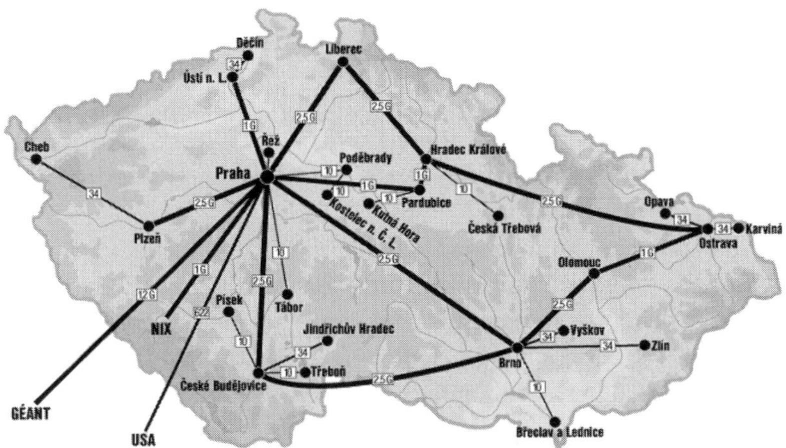

Fig. 11. CESNET.

9.8 POZNAN (Poland)

The Poznan Supercomputing and Networking Center (PSNC) has been participating in the m6bone using FreeBSD and Cisco routers. PSNC has developed its own multicast Beacon server application [26] using Java v1.4.1 with IPv6 support. Like the majority of NREN's, PSNC has obtained a

SubTLA IPv6 address space from RIPE and offers IPv6 connectivity service to its customers. PSNC has also started establishing native IPv6 connections in Poland within the POL34/155 network.

10. Summary

With the IST Fifth Framework funding drawing to a close, the focus is now on new projects to be bid for and launched within the Sixth Framework. The IPv6 Cluster and All-IPv6-World meetings have identified some key research areas. It is expected that an increasing number of projects will include IPv6 as a matter of course. With 6NET and Euro6IX delivering research infrastructure, the requirement for FP6 will be on projects like 6WINIT, demonstrating novel services and also taking a vertical view of the problem space.

The GEANT pan-European academic network is likely to support IPv6 natively (dual-stack) by mid-2003. Already some national research networks do so. As native IPv6 support in backbone networks grows more common, the requirement will shift such that those using IPv6 for daily tasks will demand a stable, predictable service. To this end, robust, production-quality links will need to be maintained to the US (e.g. Abilene) and Asia (Korea, Japan and elsewhere), and the 6bone routing "chaos" will need to be suppressed by careful introduction of route filtering and peering policies as exist in today's IPv4 networks.

As the IST projects validate and deliver new IPv6 technologies, the IPv6 Cluster will disseminate knowledge, and the IPv6 Task Forces will work with operators, vendors, researchers and governments at a national level to drive forward IPv6 adoption and deployment. In this context, collaboration and exchange with other IPv6 initiatives, including those in the US, Japan and elsewhere, will be very important. It is hoped that technology and applications can be shared and run between such networks.

Interest in IPv6 is steadily growing in the European NREN's, and knowledge and understanding of IPv6 issues is being gained as a result of the 6NET and GEANT TF-NGN activities. Full details can be found in the cited references, in particular the two GEANT deliverables. As NREN's take the first steps to dual-stack deployment, GEANT is also set to begin migration in 2003. Specific activities such as the m6bone and the Land Speed Record illustrate the capacity for NREN's to work together in furthering IPv6 deployment.

Acknowledgements

The authors would like to thank Wim Biemolt (SURFnet), Pekka Savola (Funet), Miguel Sotos (RedIRIS), Avgust Jauk (ARNES), Bartek Gajda (POZNAN), Ladislav Lhotka (CESNET) and Rina Samani (UKERNA) for contributions to this chapter.

References

[1] 6INIT Project. http://www.6init.org/.

[2] 6LINK Project. http://www.6link.org/.

[3] 6NET Project. http://www.6net.org/.

[4] 6NET Project Deliverable 2.2.1: IPv4 to IPv6 migration scoping report for organizational (NREN) networks, August 2002. http://www.6net.org/publications/deliverables/D2.2.1.pdf.

[5] 6NET Project Deliverable 2.3.1: IPv4 to IPv6 scoping report for end sites/universities, July 2002. http://www.6net.org/publications/deliverables/D2.3.1.pdf.

[6] 6NET Project Deliverable 4.1.1: Survey and evaluation of MIPv6 implementations, April 2002. http://www.6net.org/publications/deliverables/D4.1.1.pdf

[7] 6NET Project Deliverable 6.2.1: 6NET management tools requirements, July 2002. http://www.6net.org/publications/deliverables/D6.2.1.pdf.

[8] 6POWER Project. http://www.6power.org/.

[9] 6QM Project. http://www.6qm.org/.

[10] 6WINIT Project. http://www.6winit.org/.

[11] EU IPv6 Task Force Project. http://www.ipv6tf.org/.

[12] Euro6IX Deliverable D2.1: Specification of the internal network architecture of each IX. http://www.euro6ix.org/.

[13] Euro6IX Project. http://www.euro6ix.org/.

[14] European Commission Information Society Technologies Program. http://www.cordis.lu/ist/.

[15] Eurov6 Project. http://www.eurov6.org/.

[16] GEANT. http://www.dante.net/geant/.

[17] GEANT Task Force for Next Generation Networks Working Group. http://www.dante.org.uk/tf-ngn/.

[18] Internet2 Land Speed Record. http://www.internet2.edu/lsr/.

[19] IPv6 IST Cluster. http://www.ist-ipv6.org/.

[20] I. Jun-ichiro. Implementing AF-independent application.KAME Newsletter, November 1998. http://www.kame.net/newsletter/19980604/.

[21] LONG Project. http://www.ist-long.com/.

[22] SATIP6 Project. http://satip6.tilab.com/.

[23] Bermuda IPv6 Project. http://www.ipv6.ac.uk/bermuda2/.

[24] GEANT Deliverable D9.3: IPv6 testing, August 2001. http://www.dante.net/tf-ngn/D9.3.pdf.

[25] GEANT Deliverable D9.6: Report on IPv6 experiments and status, September 2002. http://www.dante.net/tf-ngn/D9.6- IPv6 test.pdf.

[26] IPv6 Multicast Beacon, developed by POZNAN, PSNC. http://dast.nlanr.net/projects/beacon.

Gigabit Network
T. Saito and H. Esaki (Eds.)
Ohmsha/IOS Press, 2003

Chapter 10

Internet2 Toward IPv6

Michael H. Lambert[a]

[a] *Pittsburgh Supercomputing Center*

Abstract. Internet2 and its backbone networks, vBNS+ and Abilene, have played a major role in the advancement of Internet research and education community in the United States. Internet2 has grown from its seminal meeting to an organization encompassing, among other efforts, engineering working groups, including one dedicated to IPv6. The Internet2 backbones, i.e., Abilene and vBNS+, have evolved considerably since their inception. Several high-performance applications have run over Internet2, setting, for the moment at least, IPv6 "land speed records". A number of topics, such as multi-homing, are of current interest to those involved with the Internet2 IPv6 effort.

1. Introduction

Internet2 [6] occupies a leading role in research and education (R&E) networking in the United States. It is a project of UCAID (University Consortium for Advanced Internet Development) and has some 200 university members in addition to government and corporate partners. A major goal of Internet2 is to facilitate high-performance networking for its members-not only "fat pipes", but also advanced technologies and services-not just for its own sake, but also as a model for the next generation of the Internet. It does not operate its own backbone network, but instead sanctions both Abilene and vBNS+ as "Internet2 backbones". It has been near the forefront of IPv6 deployment in the United States, even though one of the major motivations for IPv6, the shortage of IPv4 address space, is less of a problem in the US R&E community than almost anywhere else. The remainder of this section will present a brief history of Internet2 as well as some of its projects. This is followed by an overview of the backbone networks, with particular emphasis on IPv6 in those networks. Next follows some information on application performance on the Abilene IPv6 network, followed by a discussion of some topics of interest to the Internet2 IPv6 community.

1.1 *A Brief History of Internet2*

Internet2 came into being as a result of an October 1996 meeting in Chicago of university presidents and CIO's representing about one hundred major US research universities (R1 institutions according to the Carnegie classification [2] in use at the time). They met because of their concerns about how the commercialization of the Internet was affecting their universities' ability to conduct collaborative research and provide other services to their students, faculty and staff.

This commercialization occurred in large part because of the demise of NSFNET [9], a network which came into being as a consequence of the establishment in 1985 of several supercomputing centers funded by the National Science Foundation. The NSF model required that the majority of users access these advanced computing facilities remotely; thus a network was needed to support the centers. What arose was a collection of regional networks (such as PREPnet in the area near the Pittsburgh Supercomputing Center and NYSERnet contiguous with the Cornell Theory Center) connected via a national backbone, NSFNET.

The NSFNET backbone, provisioned by MCI and IBM and operated by MERIT and ANS, began in July 1986 as six router nodes connected via 56Kbps links in a ring topology plus one internal connection. Two years later (July 1988), the backbone consisted of thirteen core routers interconnected by several physical T1 (1.544Mbps) rings. However, the logical topology was a partial mesh of fractional-T1 (488Kbps) connections. Between July 1989 and November 1992 the network became a set of fifteen core nodes with full T1 interconnects (but without the richness of the previous logical topology). From November 1992 until its end in April 1995, the NSFNET backbone operated with thirteen core nodes interconnected at T3 (45Mbps) speed. An important point about NSFNET is that although it was created for

Fig. 1. Startup configuration of NSFNET in 1969.

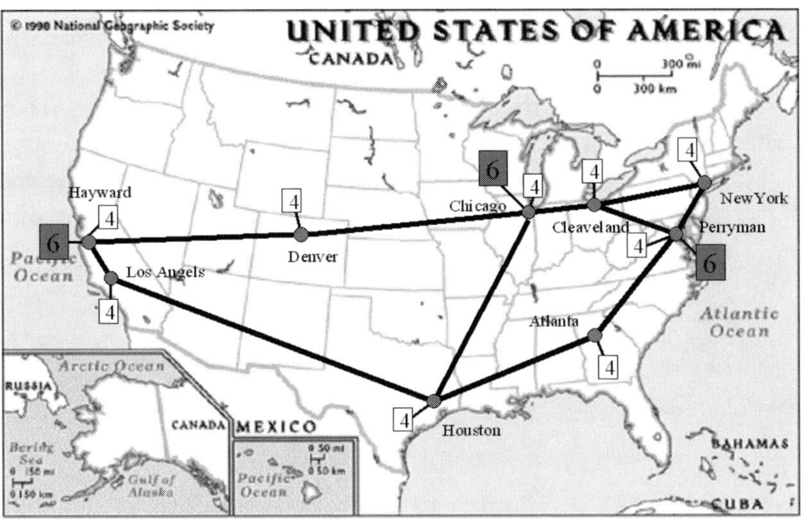

Fig. 2. vBNS as of 1998.

the purpose of serving the NSF-funded supercomputing centers, in practice it operated as a general purpose R&E network. As the commercial Internet grew, the idea of such a federally-funded network for universities became unacceptable. Faced with pressures for privatization, the NSF decided to discontinue NSFNET. To ease the transition of the regional networks to the commercial world, the NSF provided money to 1) establish Network Access Points (NAP) for public peering, 2) fund the Routing Arbiter project, from which came the route server technology, and 3) subsidize (for a short time) connectivity to the NAP's for the regional networks. At this time the NSF announced a new R&E network, to be operated by MCI (later WorldCom) under a five-year cooperative agreement. The vBNS [11] (very High-speed Backbone Network Service, Figure 2) would have one quite significant difference from NSFNET – it would only carry traffic between the NSF supercomputing centers to facilitate distributed computing. Therefore, it would in no way be a general-purpose R&E network. The NSF soon recognized that there were "meritorious applications" requiring better access to the supercomputing centers than that offered by the commercial Internet. From this need arose the Connections Program [8] that enabled universities with supercomputer users in need of high-performance networking to apply to NSF for funding to connect to the vBNS. The intent of the Connections Program was to provide vBNS connectivity only for these meritorious applications. However, it soon became apparent that the state of policy routing in 1996 was not sufficient to permit the segregation of "meritorious" and "non-meritorious" traffic. The restrictions on use of the network were relaxed, allowing general traffic between vBNS-connected institutions to transit the vBNS.

This was the state of affairs at the time of the Chicago meeting. There was a consensus that, from a technical standpoint, the vBNS was quite acceptable as a backbone network. However, other considerations, including the five year term of the vBNS cooperative agreement, led to the development of a new backbone, the Abilene network [1], another UCAID project.

1.2 Internet2 Activities

A number of activities in addition to the backbone networks fall under the aegis of Internet2. There are a number of initiatives, some of which are focused on specific constituencies (e.g., the K-20 initiative, which is chartered to bring advanced networking to educational institutions outside the major research universities) or cut a broad swath across technical disciplines (e.g., the end-to-end performance initiative, which is an Internet2 complement to the Web100 project [10]).

Internet2 has a number of working groups that fall into three broad areas: engineering, applications and middleware. Within the engineering area are working groups for multicasting, security, quality of service, routing, topology, measurement and IPv6. These working groups have a wider scope than those of the IETF and do not necessarily operate under the model of "rough consensus and running code".

The IPv6 working group [7], has, to a large degree, operated in support of the Abilene IPv6 network (which can be viewed as a shortcoming). The working group currently operates the inverse name server for the Abilene allocation. It provides input to the Abilene Technical Advisory Committee upon request. It formulated the address allocation scheme used on Abilene and makes new allocations upon request by the Abilene Network Operations Center (although registration of allocations with ARIN is handled by the NOC).

The activity of perhaps the greatest visibility and impact on the community at-large is a series of IPv6 workshops being conducted by the working group. To date, ten of these workshops have been taught around the United States, targeting network engineers at Internet2 member institutions and network aggregators. The goal is to provide them with enough hands-on experience with IPv6 addressing, exterior routing, router configuration, DNS, and basic Internet services to experiment at their home campuses and serve as a resource for their local peers.

2. Internet2 Backbone Networks

As mentioned previously, Abilene and vBNS+ (the evolution of the vBNS to include commercial customers following the end of NSF funding) are both official Internet2 backbone networks. However, because of factors to a large extent unrelated to technology, most Internet2 members are connected to Abilene and not to vBNS+. The next two sections discuss the two Internet2 backbones. Particular detail is paid to the evolution of their IPv6 services.

2.1 The vBNS+ Backbone

Deployment of the vBNS began in April 1995 following the signing of a cooperative agreement between the NSF and MCI. The original network connected five NSF-funded supercomputing centers (SCC's) at 155Mbps. This OC-3 IP/ATM network was one of the first such networks in the world. It was provisioned on MCI's commercial ATM network with most of the vBNS equipment being co-located at the SCC's. At the ATM layer were Fore ASX1000 switches. At the IP layer were Cisco 7000 routers (upgraded to 7507's in 1996) and Netstar Gigarouters (to provide HiPPI connectivity to the supercomputers).

In 1997, after the establishment of the NSF High-Performance Network Connections Program (HPNC), the vBNS backbone was expanded to include nine MCI POP's and upgraded to OC-12 (622Mbps) IP/ATM. About thirty universities, recipients of HPNC awards, connected to the network. Also in this year, IPv6 was first deployed on the vBNS. It was a purely tunneled network, at first addressed out of the original provider-based space (5f00::/8) and later migrated to the test allocation in the global, aggregatable provider-based unicast space (3ffe::/16). The first connection to this network was at the author's institution (i.e., Pittsburgh Supercomputing Center terminated on a Cisco 1005, which functioned as a single-legged router on its 10Mbps Ethernet interface (the T1 port was not used). At the time, the Cisco software for this platform even supported BGP!

The year 1998 witnessed the beginning of a backbone upgrade to OC-12 POS and Juniper M40 routers. Native IPv6 made its debut on the network: several Cisco 4700 routers were interconnected via a PVC mesh on the ATM network. The Pittsburgh Supercomputing Center moved its IPv6 connection (still via a tunnel) to a Nortel BLN router (using static routing rather than BGP).

In the years since, the vBNS (now vBNS+) has upgraded its backbone to OC-48 (2.4Gbps) POS and received one of the first production sTLA allocations from ARIN. However, it is still using the separate IPv6 backbone with Cisco 4700's. Even though it would be technically straightforward to migrate IPv6 services to run dual-stack on the core, there has been no customer demand for IPv6 connectivity at higher speeds or via media types other than PVC's or tunnels.

2.2 The Abilene Backbone

Deployment of the Abilene backbone began in 1998. The first-generation network was provisioned on fiber from Qwest Communications with Nortel Networks SONET gear and Cisco 12008 routers. NOC services are provided by Indiana University. This network consisted of nine core routers (later expanded to ten) connected by OC-48 POS trunks (some were originally OC-12 and later upgraded). Each core router had two or three connections to other core routers.

Internet2 members connecting to Abilene can do so in one of two ways: either with a direct connection to an Abilene core router or via a Gigapop; in either case, the entity connecting to Abilene is referred to as a "Connector". The primary function of a Gigapop (Figure 3) is to aggregate R&E traffic for the universities it serves; some also offer commodity Internet connectivity and other services.

Most connectors attach to the backbone at either OC-3 or OC-12 rates, using either IP/ATM or POS. In a few cases there have been connections using OC-48 POS or 1000Mbps Ethernet. The limiting factor for the latter is generally the distance between the connector router and the Abilene core router.

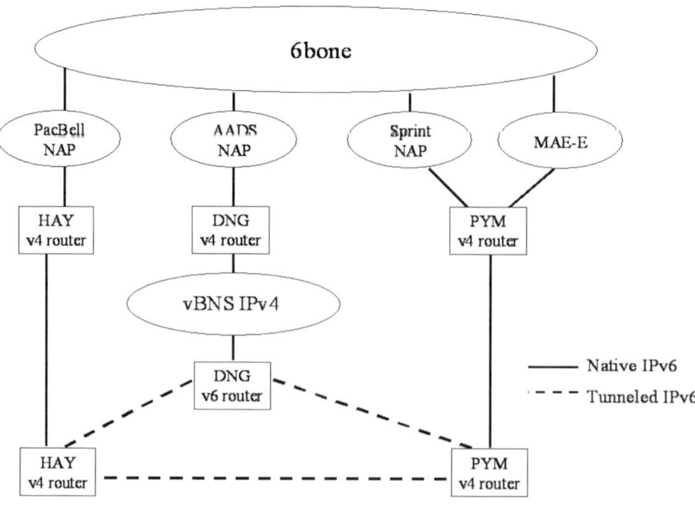

Fig. 3. vBNS Gigapop configuration.

Abilene Network Backbone - February 2002

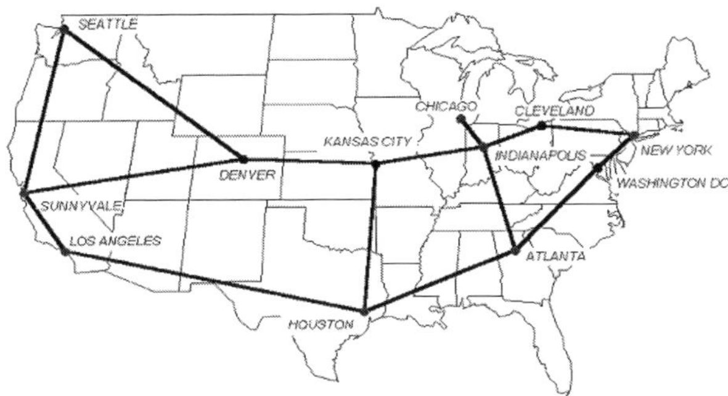

Fig. 4. Abilene backbone.

The first Abilene IPv6 backbone consisted of four Cisco 7200 routers connected by a full tunnel mesh. Three of the four routers, donated by Cisco, were located at Gigapops (in Pittsburgh, Atlanta and Boulder); the fourth was at the Qwest POP in Indianapolis, near the Abilene NOC (the Atlanta router was later moved to the Qwest POP in Sunnyvale). It was necessary to use separate routers because system software with IPv6 support did not exist at the time for the Cisco 12008 core routers. It was necessary to use tunnels because of the POS network topology. A full iBGP mesh was configured with RIPng as the IGP.

Most connections to the backbone were tunneled. However, in several cases, such as the connection to the 6TAP and WIDE (ATM) and the Gigapops hosting the routers (100Mbps Ethernet), the connections were native. During the life of this tunneled network, addressing was migrated from a 6bone pTLA to a production sTLA.

The second phase of the Abilene IPv6 backbone came during 2002, when NOC engineers determined that system software with IPv6 support was robust enough to deploy on the core 12008's to enable a dual-stack network. This deployment was performed incrementally. The first routers to be upgraded were in Indianapolis and Kansas City in order to facilitate testing between the Indiana Gigapop and Great Plains Network. Once this path proved stable, the other core routers were upgraded to run dual-stack with IS-IS as the IGP for IPv6 (OSPF remained the IGP for IPv4).

Once the backbone began running IPv6 natively, connectors and peer networks were encouraged to migrate from tunneled to native connectivity. Most did so. Because the native and tunneled networks were interconnected in Indianapolis, those connectors unable to migrate could maintain their tunneled connection. The tunneled backbone was eventually reduced to a single router.

During late 2002, the Abilene backbone (Figure 4) was in the process of being upgraded to OC-192 (9.6Gbps) links between Juniper T640 routers.

Other than the possibility of higher data rates, the only implication for IPv6 on the new network is that it was designed into the network from the beginning: IPv6 forwarding performance was a criterion in the selection of core routers. This backbone retained IS-IS as the IGP for IPv6; IPv4 was also migrated to IS-IS to take advantage of the congruent topology.

3. IPv6 Applications on Abilene

As with many IPv6-enabled networks, Abilene has had the problem of finding applications to use on the network. For quite some time, routing protocol updates were the leading traffic source on the IPv6 backbone. Recently, however, some "real" applications have been run on the Abilene backbone, showing that IPv6 networks do indeed have utility. This section will outline two of these applications.

3.1 NNTP Over IPv6

One mechanism for generating IPv6 traffic on a network is to take an existing IPv4 service, ideally a non-mission critical production service, and migrate it to IPv6. As an example of such a service, Joe St. Sauver of the University of Oregon chose USENET news, which uses NNTP as the transport protocol. For a server, a recent version of INN [5] from the Internet Software Consortium is used.

About a dozen sites from a half-dozen countries are known to be using IPv6 for NNTP transport. The University of Oregon has seen transfer rates in the 8 to 10Mbps range over the Abilene backbone. It is likely that the data rate is more of a reflection on host and application tuning than on the network itself.

3.2 DVTS

A second application which has been used successfully on the Abilene IPv6 network is the Digital Video Transport System [3]. DVTS is an application developed by the WIDE group which takes digital video, encapsulates it without compression in IPv6 (or IPv4) datagrams, and sends it across the network to be decapsulated and displayed. Given the quality of the audio-video stream, the system has only moderate hardware requirements: on both ends only a consumer-grade digital video camera with IEEE1394 capability and a mid-range computer are required.

Typical data rates for DVTS streams are 16Mbps or 32Mbps. It has been used on several occasions to facilitate remote presentations at IPv6 workshops conducted by the Internet2 IPv6 working group [4]. These presentations have taken place over both the tunneled and native networks.

4. IPv6 Working Group Activities

As with many entities of its type, only a small number of the members of the Internet2 IPv6 working group are actively involved in its activities. Thus, much less can be accomplished than many would desire.

Efforts to foster the deployment of IPv6 in the US R&E community have been moderately successful at best to date. Many campuses have IPv6 connectivity as a result of workshops, but in most cases these extend to only a few workstations. Because IPv6 is viewed as just a gradual evolution of IPv4, and the networking staff at many institutions have major constraints on their time, there has been no strong push for IPv6 deployment. A killer application for IPv6 would likely change that.

Another topic which is of current interest to the working group (not to mention the rest of the global IPv6 community) is the problem of multihoming. Under discussion are such items as what prefix lengths to accept from peers and whether or not to accept non-Abilene prefixes from connectors. The latter is relevant because some Gigapops now have their own allocations from ARIN – should Abilene require them to announce only Abilene-allocated prefixes?

Multiple provider-aggregated addresses on an interface are especially problematic in the Abilene environment. When dealing only with commodity connections, all addresses are more or less equal. However, because of potential conditions of use on R&E networks, addresses can have semantics which are hidden from the end host. It may take some time before a workable solution is found.

5. Summary

Internet2 and its member universities have seen the IPv6 backbone available to them grow from a set of ad hoc tunnels to one designed to forward IPv6 traffic at speed up to 1Gbps. It has tried to foster the deployment of IPv6 through workshops intended to increase the awareness of the protocol in the Gigapops and universities. Finally, Internet2 has demonstrated at least moderately high-performance applications over IPv6.

 10. Internet2 Toward IPv6

References

[1] Abilene network operations center. http://www.abilene.iu.edu.

[2] Carnegie classification of institutions of higher education. http://www.carnegiefoundation.org/Classification.

[3] DV (digital video) over IP (DVTS). http://www.sfc.wide.ad.jp/DVTS.

[4] Firing up DVTS over IPv6. http://www.internet2.edu/ipv6/presentations/DVTS-Presentation.ppt.

[5] Internet software consortium. ftp://ftp.isc.org.

[6] Internet2: Internet2 website. http://www.internet2.edu.

[7] IPv6 WG homepage. http://ipv6.internet2.edu.

[8] NSF connections program slides. http://moat.nlanr.net/INFRA/Luker/index.html.

[9] NSFNET. http://moar.nlanar.net/INFRA/NSFNET.html.

[10] TheWeb100 project. http://www.web100.org.

[11] Welcome to vBNS. http://www.vbns.net.

Gigabit Network
T. Saito and H. Esaki (Eds.)
Ohmsha/IOS Press, 2003

Chapter 11

AI3 Satellite Internet Infrastructure in Asia

Tomomitsu Baba[a] and Suguru Yamaguchi[b]

[a] Kurashiki University of Science and the Arts;
[b] Nara Institute of Science and Technology

Abstract. The Internet has become a critical communication infrastructure in many countries. In Asia, a strong demand for Internet development persists because only few people have unfettered access. The Asian Internet Interconnection Initiative (AI3) was established in 1995 by the WIDE Project to develop technologies for accelerating Internet development in Asia. Since 1995, the project has been constructing its own testbed network using satellite communication systems, making it the only such satellite-based international R&D network in the world. Currently this network connects sixteen member sites in ten countries.

Our activities include engineering an infrastructure using Ku band satellite transponders for Internet applications, UDL routing for asymmetric one-way links to maximize available bandwidth from hub stations to member sites, and a series of experiments including remote education to confirm system usability.

In this report, we give an overview of the AI3 project's achievements and describe its challenges for the future.

1. Introduction

The Internet has become a critical part of the infrastructure for current society. It provides a variety of services that many people already cannot imagine living without. Because this level of penetration has taken five or more years in developed countries, it is natural to assume that the process will take even longer in developing ones. In fact, many developing countries are now struggling to develop their Internet infrastructure; however, the number of people who can access the Internet without difficulty in these countries is still low. The governments in many developing countries are now encouraging industry and other sectors of society to help resolve this "digital divide", but significant progress has yet to be made in most countries. Stagnation in many areas has been the result. Many international projects have tackled this problem since the mid 1990's. For example, the Internet

Society has been actively working on human resource development through organizing Networking Training Workshops (NTW) [1] since 1992. The workshop gathers people involved in Internet development from various fields and provides in-depth training for Internet development. This workshop has contributed significantly to Internet development, especially in Latin America and African countries. Also, many international donor programs such as JICA in Japan have been working aggressively on Internet development in various countries. Of course, there has been tremendous commercial investment for Internet development as well.

Even with these activities, however, a large gap still remains between developed and developing countries in terms of Internet development. For Asian countries, especially, this penetration process has been made more difficult by large gaps in economic development, the existence of a tremendous number of languages, and a geographical and climatic diversity that ranges from small tropical islands in the Pacific Ocean to monsoon regions in Southeast Asia, deserts in Central Asia, and ice-bound rural areas in the north. In the mid 1990's, leaders involved in Internet development in Asia concluded that more active participation in Internet development was an absolute requirement for success.

It was against this background that the Asian Internet Interconnection Initiatives, or AI3 (ei-tripl-ai), was established in 1995 to work toward Internet development in Asia [2,15,16]. When we began this project, we assumed the following conditions for accelerating the process of Internet deployment: (1) a testbed network for live demonstrations of the technology, since such a showcase for the technology would be effective in persuading people of its potential, (2) research for regional adaptation and localization of the Internet should be conducted simultaneously with deployment, and (3) local human resource development to produce supporters and participants is vital for rapid deployment.

With these assumptions in place, the AI3 project decided to start as a research consortium of leading research groups in Asian universities. Because universities are charged with human resource development, have fewer restrictions on maintaining a testbed network, and have a base of their own research activities, we expect to find many researchers working actively on Internet technologies.

In the seven years since its inception, the AI3 testbed network has connected to sixteen universsities in ten countries in the region and is still expanding. The network has been operating around the clock and has become the default communication infrastructure for members of the AI3 project. In this report, we summarize the AI3 project and its achievements in Internet development as well as our R&D process using the AI3 satellite Internet infrastructure in Asia.

2. Infrastructure

2.1 Layer 2 (L2) design

When we started the AI3 project in the mid 1990's, there was fiber-optic infrastructure in Japan; however, legacy PSTN was only the major infrastructure available in many Asian countries. Therefore, our first challenge was finding a way to establish an Internet infrastructure for our members in Asia despite a lack of terrestrial communication infrastructure. The solution we decided upon was to build our international Internet testbed network using communication satellites.

The advantages of using communication satellites as L2 technology are significant. Many of our members do not have adequate terrestrial infrastructure to establish broadband Internet connections, but the only infrastructure requirement for the construction of satellite ground stations are adequate and stable electricity.

Since technology transfer and equal partnerships are important in any international collaboration, the AI3 project crafted a project framework stating that the purchase of VSAT ground stations, obtaining appropriate licenses from the local authorities, and all other activities requiring local coordination must be completed by each partner. However, the AI3 project does provide satellite circuits (bandwidth on transponders) and related technical information to all partners. Within this framework, AI3 partners were invited from Asian countries, and groups who demonstrated sufficient need for the AI3 project were selected.

The AI3 project provides a base level of aid for the construction of satellite ground stations, including the minimum requirements for station equipment including frequency and polarization devices, satellite modems, antennae, and HPA (High Power Amplifier). Along with these minimum requirement, AI3 partners set up their ground stations considering their own procurements costs, maintenance, and operation needs.

Consequently, each ground station was set up using different RF units, as shown in Figure 1. Through this process, we found that there were various regulations governing satellite communications in Asian countries. Licenses were required from local authorities for wireless communications, ground station location, import permissions for wireless devices, and other project elements. Obtaining these licenses was harder in some locations than in others, but ultimately all AI3 partners were successful. Thus, from the aspect of human resource development, all AI3 partners received a very educational experience from the very first stage of the project.

Since the early stages of the AI3 project, Ku band transponders on the communication satellite JCSAT-3, operated by JSAT Inc., have been utilized.

Fig. 1. AI3 ground stations using different RF units.

Table 1. AI3 partners connected via bi-directional link on both Ku and C bands

Member	Country	Abbreviation
Ku band		
Nara Institute of Science and Technology	Japan	NAIST
Institute of Technology Bandung	Indonesia	ITB
Hong Kong University of Science and Technology	China	HKUST
Asian Institute of Technology	Thailand	AIT
C band		
Keio University	Japan	KEIO
Temasek Polytechnic	Singapore	TP
University Sains Malaysia	Malaysia	USM
Advanced Science and Technology Institute	Philippines	ASTI
Institute of Information Technology	Vietnam	IOIT
University of Colombo	Sri Lanka	CMB

Ku band VSAT ground stations were installed at the member sites listed in the upper part of Table 1 around 1996. At that time, there existed much hope for the possibility of using the Ku band satellite communication channel for Internet traffic because there had been a few experiments using Ku band, and it shows a greater rain degradation effect than C band, which is quite popular for digital communication in tropical region. Therefore, several experiments to confirm usability of Ku band satellite communication channels in tropical areas, which experience frequent heavy rains, were conducted immediately after our installation. Furthermore, when we began this project, there was no other large-scale satellite Internet infrastructure existent, so we developed the system and tools to operate our infrastructure to be merged onto the ordinary Internet environment.

In 1999, we had an opportunity to expand our activities to more countries in Asia. At the time, we started using C band transponder on JCSAT-3 and

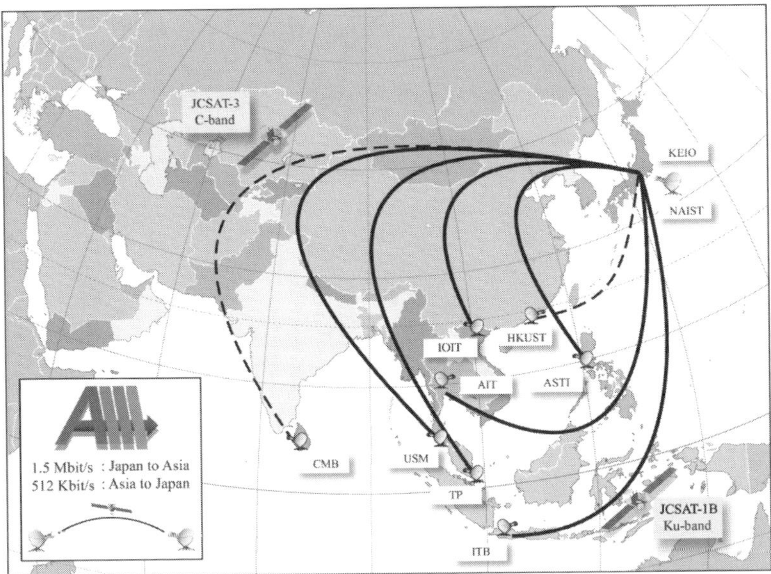

Fig. 2. Testbed network.

added five more universities, which are listed in the lower part of Table 1. These universities added to our C band infrastructure are relatively strong in terms of technology development in the region. In Figure 2, our network is mapped onto geographical points, though two of them are not in operation due to operational difficulties at HKUST stations and planning of a link to CMB before installation.

With two transponders, for both Ku and C bands, we installed two hub stations in Japan to connect to our member sites, and set up broadband transit links between these stations as illustrated in Figure 3 using the nation-wide ATM network JGN [9] in Japan.

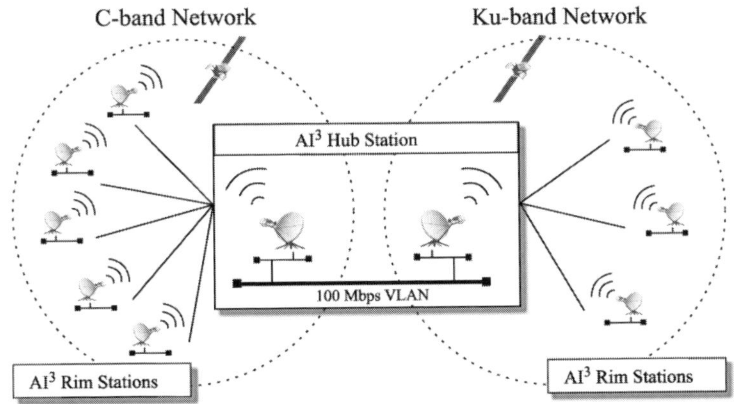

Fig. 3. Network junction between Ku and C bands.

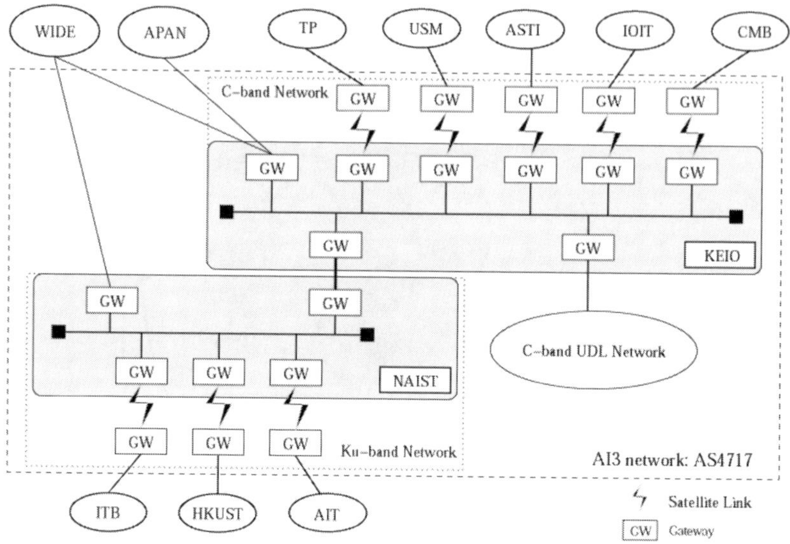

Fig. 4. Layer 3 logical map.

2.2 Layer 3 (L3) design

Our network consists of multiple AS's residing on member sites. Figure 4 shows a Layer 3 logical map of this infrastructure. We operate our own AS for the project, and our network exchanges traffic to and from our member AS's. In this sense, our network can be considered a distributed L3 internet exchange ranging over the southeast Asian region.

For global Internet connectivity, APAN TRANSPAC link [3,10] as well as WIDE Internet operated by the WIDE Project, can be used. This broadband backbone Internet connectivity facilitates experiments with other members in the AI3 project as well as research groups in Internet2 and other next generation Internet consortia. Our network is especially well integrated into APAN, as shown in Figure 5.

2.3 UDL (Uni-Directional Link)

One of the project's major achievements is UDLR, or Uni-Directional Link Routing [7].

In the original design for VSAT ground stations, each station had to obtain appropriate licenses from authorities to receive and transmit communication satellite signals. However, as described above, such signal transmission usually requires a long and complicated application process. Moreover, VSAT ground stations with transmission capability require expensive equipment. These characteristics of VSAT ground stations made our licensing process

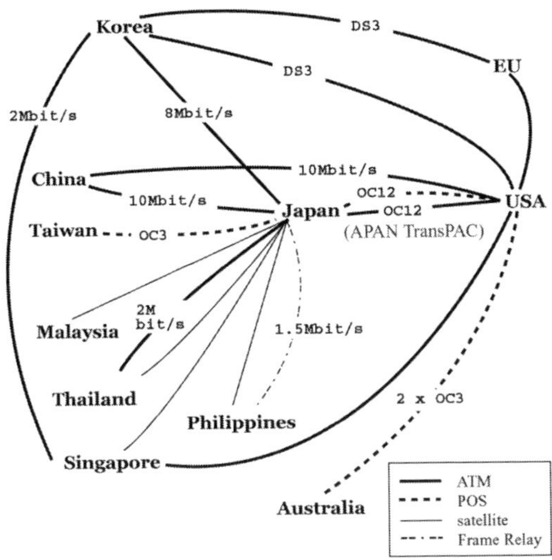

Fig. 5. AI3 network fully integrated into APAN and other international R&D networks.

troublesome in many cases, making it in many cases the largest hurdle for our project members.

On the other hand, reception-only ground stations are quite popular in Asian countries, especially for TV broadcasting, and in many cases there are no licensing requirements for building reception-only ground stations. Furthermore, reception-only stations are inexpensive. Using them for Internet connectivity will ease the process of installing gateways at our member sites.

Around 1998, the basic technology for UDL routing was developed. As shown in Figure 6, a single satellite channel is shared among multiple UDL receivers. The "feed" station transmits IP datagrams to the satellite channel as an L1 broadcast, and each UDL receiver selectively receives and forwards IP datagrams bound for the network connected to the receiver. For uplinks from UDL receiver sites to the Internet, terrestrial links from the site to the Internet are utilized. Of course, many Internet protocols are based on the assumption that each Internet link is bidirectional, so the UDL receivers and the feed station must be able to handle routing control to make UDL work consistently with other connected networks. Details of UDL operation are available in the references [4].

This mechanism fits well with both communication satellites and the many usages of the Internet. Since a single communication satellite may be considered an amplifier, receiving signals from ground stations and retransmitting them to areas covered by its transponder, the nature of conducing communications via satellite is akin to broadcasting. With UDL routing technology, we must share satellite bandwidth with other members, but theoretically the bandwidth of an entire single transponder may be used,

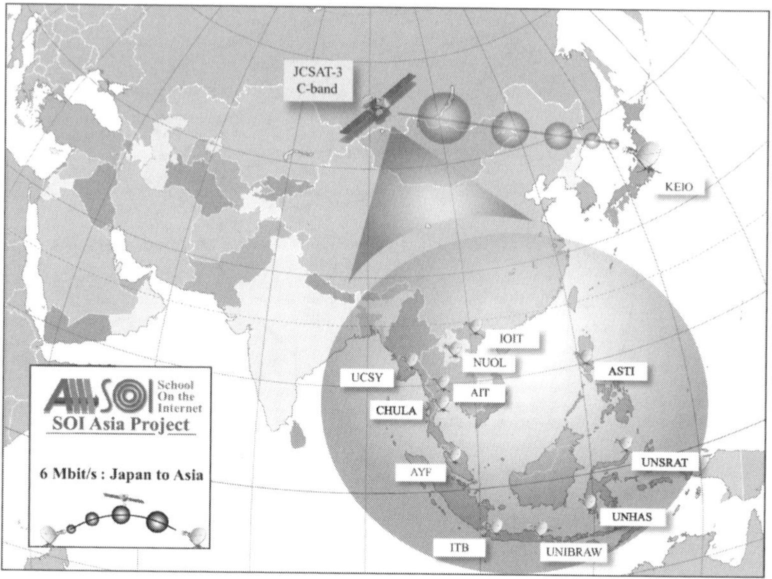

Fig. 6. UDL network in AI3.

which is up to 30Mbps for a one way link. (Current design of communication satellites is such that a single satellite has multiple transponders. A single transponder for Ku and C bands can provide bandwidth of around 30Mbps. Of course, in other frequency band such as S or Ka, the bandwidth provided by a single transponder is varied and not limited to 30Mbps.) Currently we are providing 6Mbps for UDL sites, which is sufficient for multimedia traffic, such as remote education using teleconferencing, and other ordinary Internet applications.

In the UDL configuration, we must use terrestrial links to uplink to the Internet. Available bandwidth on these terrestrial links in the AI3 project is usually about 64Kbps. However, since much usage of the Internet still follows the client-server model, clients residing in a UDL receiver site must transmit data destined for application servers on the Internet even though bandwidth is quite limited. In our experiments, 64Kbps uplink is sufficient for ordinary usage in many cases. With this UDL technology, eight more universities in Malaysia, Indonesia, Laos, and Myanmar are now connected.

All project members are listed in Table 2.

3. Technology and Application Development

We have been developing and evaluating new technologies and applications among AI3 partners over the AI3 satellite Internet infrastructure.

Table 2. AI3 partners using UDL

Member	Country	Abbreviation
Keio University	Japan	KEIO
Chulalongkorn University	Thailand	CHULA
Asian Institute of Technology	Thailand	AIT
National University of Laos	Laos	NUOL
University of Computer Studies, Yangon	Myanmar	UCSY
Brawijaya University	Indonesia	UNIBRAW
Sam Ratulangi University	Indonesia	UNSRAT
Hasanuddin University	Indonesia	UNHAS
Institute of Technology Bandung	Indonesia	ITB
Asian Youth Fellowship	Malaysia	AFY
Institute of Information Technology	Vietnam	IOIT
Advanced Science and Technology Institute	Philippines	ASTI

Network management is important to keep our infrastructure operating optimally. Unlike the ordinary Internet system, we are using satellite-based IP links for our backbone, which creates the need for a satellite Internet monitoring system. With our satellite Internet infrastructure, we face many difficulties in daily network operation. For example, because the network is composed of various communication segments such as IDU and ODU, and is spread over multiple countries, its complexity and dispersed nature make it difficult to analyze a fault point when a link is down. We also face difficulties in monitoring normal status without constant records, since although we are familiar with Internet technology, we do not have the same level of expertise regarding satellite communication system.

In general, electronic devices consist of ODU and IDU in VSAT ground stations, as illustrated in Figure 7.

Typical data obtained from ODU is related to air temperature and precipitation, while some IDU models allow for more weather-related data

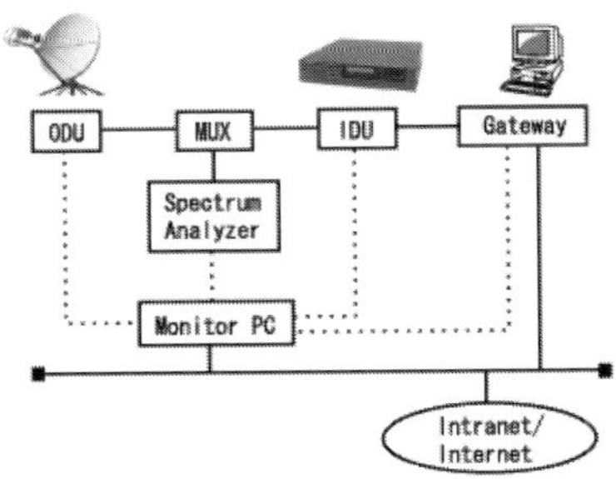

Fig. 7. System components of AI3 satellite internet monitoring.

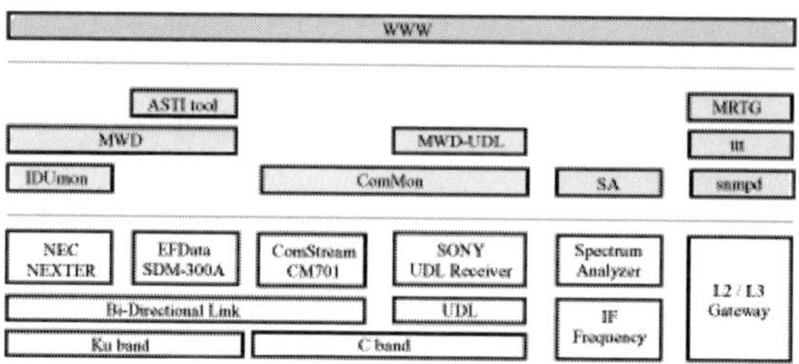

Fig. 8. Web-based satellite internet monitoring system.

collection. IDU such as satellite modems supply fluctuation information such as C/N (Carrier over Noise Ratio) and AGC (Automatic Gain Control) level. Spectrum Analyzer, which monitors IF frequency at the hub station, allows receiving signal power from the rim stations to be changed. Furthermore, Layer 2 and Layer 3 gateways show the link status, IP traffic flow, protocol distribution, and other L3 related statistical data. These components make it easier to diagnose satellite Internet infrastructure.

Figure 8 shows a web-based monitoring system. This figure is divided into three layers. The system components to be monitored, such as satellite modems, are placed at the lowest layer. The product names of our monitoring software appear at the middle layer. In this layer, the software we are using, such as MWD, (Modem Watch Dog) is placed corresponding to each monitoring target. For example, MWD supports monitoring of the two satellite modems NEC NEXTER and EFData SDM-300A. These products have been developed by AI3 partners because of the necessity for trouble-shooting and data measurement. These monitoring results are shared among the members through the website as shown in the upper portion of the figure. Using these monitoring tools, we have already started collection of statistical data at some sites because such statistical information is helpful in diagnosing the condition of the infrastructure.

The AI3 Philippines group has deployed weathermap software which visualizes all IP traffic flow in the AI3 testbed network. This information can then be viewed by other international R&D communities [14].

AI3 Cache Bone has been developed as a hierarchical WWW cache using multicast over satellite for cache delivery. This system is to provide both higher hit rates and saving of overall bandwidth consumption through fetching WWW objects and multicasting them over satellites [5,6,13]. Figure 9 shows a basic model of AI3 Cache Bone. In the Asian region, we confirmed that the performance of our rim cache is much improved using our bi-directional satellite link when compared with an ordinary hierarchical cache system such as Squid. All system implementation, including the multicast push caching, has been finished, and further deployment of AI3 Cache Bone over our UDL network is expected.

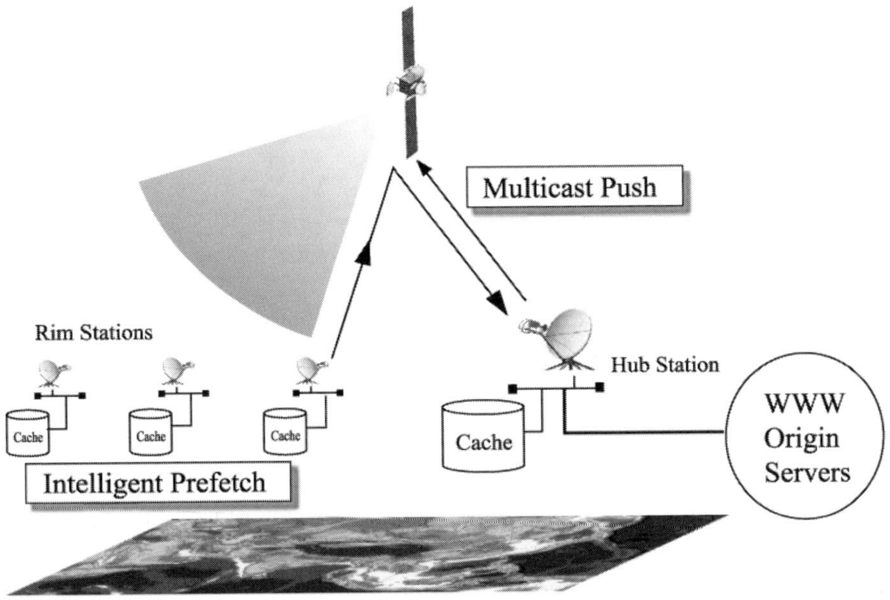

Fig. 9. AI3 cache bone.

IPv6 is a key technologies for the next generation Internet. In the AI3 testbed network, 6bone-AI3 is in operation and is connected with 6bone. We are gradually expanding IPv6 experimental networks to Asian countries through our testbed network. Our deployment efforts have accelerated IPv6 promotion and regional deployment thus far in Asian countries such as Indonesia, Philippines and Malaysia. For example, in Malaysia, a Multimedia Conferencing System called "MCS" has been developed by USM, AI3 Malaysia group[13], which began the work of enabling MCS to run as an IPv6 application.

4. Human Resource Development

There is still strong demand for human resource development for Internet infrastructure in Asia. In AI3, we have designed several activities to focus on human resource development.

Since AI3 can be defined as a collection of leading institutions on Internet technology in the Asian region, each member has its own capability of nurturing human resources in the area of Internet technology, engineering and operation. For many members, however, some leading edge technologies such as distributed WWW cache and multimedia applications were beyond their capabilities when we started this project. Therefore, technology transfer was a high priority in the early days.

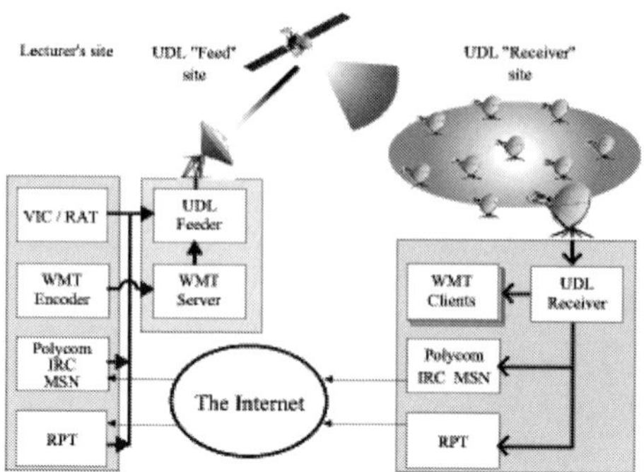

Fig. 10. SOI Asia application over UDL network in AI3.

In 2000, along with UDL deployment, we started a project called "SOI Asia", or School of Internet in Asia. SOI Asia provides lectures on technologies needed to continue the project, but also offers more lectures on various topics for higher educations at universities in the Asian region [11]. With UDL, we've been able to offer lectures to a greater audience as well as stored (recorded) sessions in SOI Asia program with less overhead. Currently, many lectures provided in this program are utilized at our member universities, even at sites with a 9.6Kbps PPP uplink to the Internet.

Figure 10 shows that some live lecture applications with multicast capability are used in the SOI Asia program [11]. Lecturer's sites are connected to a UDL feed site with enough bandwidth. The UDL feed site is located at Keio University in Japan. WMT (Windows Media Technology) components are used to stream the live lectures. Some interactive communication tools such as VIC/RAT are used for live Q&A sessions with less delay than WMT. Lecture materials like Power Point are synchronized using RPT software developed by the SOI project[8].

5. Future of the AI3 Project

Seven years have now passed since we started this project in 1995. Through our activities, several issues have been raised as challenges for the future.

5.1 Combination with Terrestrial Broadband Cables

Asian countries are still the target of huge investment for international optical fiber cable installations. Recently, several cable consortiums announced their plan to install cables from US, Japan, Australia, and even Europe to Asian countries within five years. Destinations for these cables are normally coastal cities close to economic centers, while domestic infrastructures in some countries is being improved to provide more Internet services at the national level. Using these cables would allow us to implement broadband connectivity with some members in the AI3 project.

The reason why we started this project with communication satellite channels was that there was no cable infrastructure around many members of the AI3 project. In fact, in the early stages of the AI3 project, many of our members were islands connected to the Internet via the satellite channel provided by the AI3 project, while connectivity to other domestic Internet sites remained through narrowband connections such as 64Kbps links. If our link between our sites in Japan and member sites can be upgraded to a broadband level of up to several hundred Mbps, we can transform our member sites into hubs for more activity. For example, remote education efforts using our UDL infrastructure in the AI3 network has a bottleneck of bandwidth availability from member sites to UDL feed sites in Japan. While it is quite easy to deliver lectures from our feed sites in Japan, it is also necessary to have enough bandwidth to go in the other direction from the member sites back to the Japanese feed site. In ordinary situations, at least 256Kbps should be available to enable lectures with video and audio transmission and manage real time conversation in Q&A sessions. This requirement is a substantial hurdle to conducting lectures from our member sites. Once we have broadband connections between sites in Japan and at least one of our members in southeast Asia, we can make that member a hub with a full functional satellite ground station to work as a UDL feed site, which can then provide its own remote education directly with less overhead. Also, if the domestic link to the new hub site is well equipped and offers sufficient bandwidth, other sites around the new hub will be able to deliver remote educations programs through the hub as well. If the hub is hooked up to satellites other than JCSAT-3, then obviously our network coverage can be changed to cover other areas of Asia.

We have just begun a feasibility study on installing a broadband connection to one of our member sites. At present, the study is still ongoing; however, it does promise significant potential for broadening our activities in Asia.

5.2 Working with Industry

Through the AI3 project, there have been several technologies developed which have great potential accelerate Internet deployment. However, for larger scale deployment, participation from industry or other sectors is vital for turning these technologies into products and services. Specifically, there is such strong demand for remote education in Asia and the Pacific region that many in our project are expecting to expand our efforts using UDL. However, it can still cover the currently limited number of subscribers and areas of interest. In order to provide greater coverage, it is hardly a trivial task to transfer these technologies to industry or other sectors to commercialize them into a sustainable service. Of course, this process would also have to include localization to adapt technologies to each country in order to achieve a greater coverage area. In developed countries such as Japan, it would be possible to either commercialize these technologies or ask non-profit organizations to operate them, but we have the impression that it would be difficult to achieve this industrialization in sustainable fashion in developing countries. However, making steps forward in providing our technologies to accelerate Internet deployment in Asia has been our project's commitment since its inception. Therefore, we must develop a strategy and create a feasible plan to help the technology we have developed take root in the Asian region.

6. Summary

In this report, we have provided an overview of the AI3 project, which was established in 1995 by the WIDE project to aid development of technologies for accelerating Internet deployment in Asia. Currently, the project is working actively with seventeen members from eleven countries in the region. In last seven years, this project has contributed to the development of technologies such as UDL while pursuing challenges such as the remote education program SOI-Asia on the AI3 satellite infrastructure. Seven years have passed and now we are facing new issues: development of new architecture using both communications satellite and terrestrial optical fiber cables to enhance our coverage in Asia and industrialization of our technologies in order to transform them into a sustainable services. Several new activities are now underway as a means of meeting the challenges that these new issues present.

References

[1] Network Training Workshop 2000. http://www.isoc.org/inet2000/ntw.shtml.

[2] T. Baba, H. Izumiyama, and S. Yamaguchi, "AI3 Satellite Internet Infrastructure and the Deployment in Asia", IEICE Trans. Commun. Vol. E84-B, No. 8, pp. 2048–2057, August 2001.

[3] K. Chon, "Global High Performance Research Network: an Asia-Pacific Perspective", Proceedings of WWCA'98, March 1998.

[4] W. D. E. Duros, H. Izumiyama, N. Fujii, and Y. Zhang, "A link layer tunneling mechanism for unidirectional links", RFC 3077, March 2001.

[5] H. Inoue, K. Kanchanasut, and S. Yamaguchi, "An adaptive WWW cache mechanism in the AI3 network", Proceedings of INET'97, June 1997.

[6] H. Inoue, T. Sakamoto, K. Kanchanasut, S. Yamaguchi, and Y. Oie, "Webhint: An automatic configuration mechanism for optimizing World Wide Web cache system utilization", Proceedings of INET'98, June 1998.

[7] H. Izumiyama and H. Kusumoto, "Implementation and operation of satellite based broadcast IP network", SAINT2003, January 2003.

[8] K. Kawai, K. Okawa, and J. Murai, "Practical Experiences of Higher Education on the Internet – Cases from the School of Internet", Proc. of ICCC'99, September 1999.

[9] K. Kobayashi, S. Katsuno, K. Nakamura, Y. Mikamo, H. Hayashi, A. Machizawa, Y. Kitatsuji, and H. Esaki, "JGN IPv6 Network", SAINT2003, January 2003.

[10] K. Konishi, K. Chon, and S. Goto, "APAN: Asia-Pacific advanced network", Proceedings of PTC2000, January 2000.

[11] S. Mikawa, K. Okawa, and J. Murai, "Establishment of a lecture environment using Internet technology via satellite communication in Asian countries", SAINT2003, January2003.

[12] R. Sureswaran, B. Rahmat, and V. Siva, "Satellite Bandwidth Requirements for an MCS Based Multimedia Conference", SAINT2003, January 2003.

[13] P. Tantatsanawong, H. Phien, and K. Kanchanasut, "Modeling and forecasting of hourly transactions on a WWW and proxy cache server", Proceedings of Internet Workshop'99, February 1999.

[14] E. D. Vinas, C. P. Quiblat, C. M. R. Camus, R. C. Mendoza, D. F. Villorente, and T. Baba, "Characterization, Analysis, and Visualization of Traffic in the Asian Internet Intercon-nection Initiatives (AI3) Satellite-based Research Network Test Bed", SAINT2003, January 2003.

[15] S. Yamaguchi, H. Izumiyama, and J. Murai, "Sustainable Collaborative Efforts in Internet Development in Asia: AI3 Phase II", Proceedings of INET'97, June 1997.

[16] S. Yamaguchi and J. Murai, WAsian internet interconnection.

Author Index